Mathematics *and* Religion

TEMPLETON SCIENCE AND RELIGION SERIES

In our fast-paced and high-tech era, when visual information seems so dominant, the need for short and compelling books has increased. This conciseness and convenience is the goal of the Templeton Science and Religion Series. We have commissioned scientists in a range of fields to distill their experience and knowledge into a brief tour of their specialties. They are writing for a general audience, readers with interests in the sciences or the humanities, which includes religion and theology. The relationship between science and religion has been likened to four types of doorways. The first two enter a realm of "conflict" or "separation" between these two views of life and the world. The next two doorways, however, open to a world of "interaction" or "harmony" between science and religion. We have asked our authors to enter these latter doorways to judge the possibilities. They begin with their sciences and, in aiming to address religion, return with a wide variety of critical viewpoints. We hope these short books open intellectual doors of every kind to readers of all backgrounds.

Series Editors: J. Wentzel van Huyssteen & Khalil Chamcham
Project Editor: Larry Witham

Mathematics
and Religion

OUR LANGUAGES OF SIGN AND SYMBOL

Javier Leach

TEMPLETON PRESS

Templeton Press
300 Conshohocken State Road, Suite 550
West Conshohocken, PA 19428
www.templetonpress.org

Typeset and designed by Gopa & Ted2, Inc.

Library of Congress Cataloging-in-Publication Data
Leach, Javier, 1942–
  Mathematics and religion : our languages of sign and symbol /
Javier Leach.
    p. cm.
  Includes bibliographical references and index.
  ISBN-13: 978-1-59947-149-5 (pbk. : alk. paper)
  ISBN-10: 1-59947-149-3 (pbk. : alk. paper) 1. Mathematics.
2. Religion and science. 3. Signs and symbols. I. Title.
  BL265.M3L43 2010
  201'.6510—dc22
                                                    2010008268

Printed in the United States of America

10 11 12 13 14 15   10 9 8 7 6 5 4 3 2 1

Contents

vi : CONTENTS

Preface

AT ANY OF the world's great tourist sites—Paris, Tokyo, or Mexico City, for example—we typically see travelers using dictionaries to translate their native tongues into the local language, reminding us that our natural languages still divide us. At the same time, a language that seems to unite us wherever we go is the language of mathematics. Whether we are traveling in Germany, Indonesia, or Brazil, we can do business together because we agree that 2 + 2 = 4. The mathematical laws of gravity and the three dimensions of space allow us to find directions to New York's Central Park or fly home, over a curved planet, on a jetliner.

This book is about our languages, but with a focus on the privileged role that mathematics has in our ability to communicate about the world around us. In this sense, mathematics is our public language, but it is more than that also. As I hope you will see, mathematics leads us through science and brings us to questions about a greater reality called metaphysical reality, which we usually approach in the context of philosophy and religion.

For many years, my friends, students, and acquaintances have asked me how I—a university professor of mathematics, logic, and computer science—reconcile my profession with my other vocation as a Jesuit priest. How does a person reconcile science and religion, both intellectually and on a personal basis? This book on mathematics and religion has also allowed me to explore that question further.

Mathematics is a difficult language for the uninitiated, especially when we reach beyond elementary arithmetic, geometry, and algebra to what I call the modern formal language of mathematics—the language we apply, for example, to computer science. In this advanced realm of math, we also probe the sufficiency of mathematical systems themselves. Do these systems have the power to explain an ultimate reality, or are they more like tools we invent to simply do the job: to measure the construction of a bridge, send a rocket to the moon, or run a computer calculation? By pursuing these advanced areas of mathematics, we finally arrive at questions about the truth and consistency of any system, whether a scientific system that talks about nature or a theological system that talks about God.

I have organized this book in the following way. The first chapters focus on defining and explaining the three basic languages that concern us. Each has a particular kind of perception of reality and then a system of signs or symbols (a language) to convey that perception.

The first kind of perception is found in logic and mathematics, a purely mental kind of perception. It uses the language of formal signs. This language often seems to be an array of impenetrable hieroglyphics, such as the notation $\Sigma ar = (0, 1, +, \times, <)$. But, as we will see, everyone can understand this language to some degree. Of course, the language is also very specialized, the lingua franca of logicians and mathematicians.

The second kind of perception of the world is through empirical science, which uses representational language to convey that perception. This language speaks of physical realities such as weight, force, or mass, and is the language of physics, chemistry, geology, and neuroscience. Famously, for example, Newton gave us a representational language to talk about the force of gravity based on the mass and distance between objects such as planets. Einstein did likewise with a language that says energy equals mass in the particular arrangement of $E = mc^2$.

Finally, a third kind of human perception and language is metaphysical. It is a logical language just like mathematics, but it uses symbols (not signs) to explain perceptions of relationships, causes, and ultimate reality. These symbols have included the idea of a transcendent God, a being that surpasses all other beings, or a being that is in relation to the world, but is nevertheless beyond the world. Metaphysical language uses symbols to speak about the individuality and unity of things, the nature of the infinite, the scope of the universe, or the relationships we call community.

It took most of human history for us to arrive at our modern understanding of mathematical language. The largest section of this book is a survey of the evolution of our mathematical systems, which I find the best way to introduce mathematics to a general audience. In hindsight, we can now see the turning points in our growing knowledge of mathematics and reflect on the colorful stories of the people who moved the science of mathematics up to its present state. As we will see, the formalization of mathematical systems characterizes that current state.

After the history, we look at two of the most fundamental mathematical languages: propositional logic and first-order predicate logic. Please don't let these names—which we technically call L0 and L1—make the section on propositional logic (chapter 7) offputting. This is fairly technical material, but I have tried to explain it in a narrative of natural language and have added accompanying appendices for students who wish to pursue this topic more extensively. For a general reader, this introduction to the formal language of logic offers a useful overview and a glimpse at how mathematicians and computer experts think and talk today.

The book concludes with a look at how we derive meaning from the semantics of our various languages—mathematical, empirical, and religious—and in what ways metaphysical questions have become more important as our culture grows more scientific. Our ability to understand our various kinds of languages helps us not only to take advantage of a complex scientific world but also to deepen our search for personal answers to the great metaphysical questions.

Personal experience often motivates the desire to embark on such metaphysical search. The crucial experience in my life came in my homeland of Spain, where I began as a philosophy student. I later pursued advanced studies in mathematics, attracted by its beauty, clarity, and technical power. I had been studying mathematics at the university for three years when, in May 1968, the student revolution spread the euphoria of cultural change across Western Europe. Traditional values such as religion, patriotism, and respect for authority were called into question. Equality, sexual liberation, and human rights were affirmed. This was also the decade of the Second Vatican Council (1962–65). Vatican II represented an effort to open the Catholic Church to the current culture on several fronts. The outcomes of this effort included: updating liturgical language, incorporating human rights values into the life of the church, recognizing the basic equality of all baptized, and acknowledging the right to religious freedom and the need to cooperate with other religions.

My own questioning in this period led me to study theology, beginning in the early 1970s in Frankfurt, Germany. In my own country, where I returned to teach, the long rule of General Franco was coming to an end with his death in 1975. The political transition to democracy was a major cultural change in Spain. Having entered the priesthood in these years, I also continued as a professional mathematician. Today I teach logic and mathematics in the department of computing at the Complutense University of Madrid, one of the main public universities in Spain.

In the past three decades, the world has seemed to change more rapidly than ever. Math and science have grown in importance. To the surprise of many, we also see a perennial return to religion. In such times, I believe that mathematics retains a privileged position because of its unique role in linking—by the principles of logic— science with philosophy and theology. In making this case, I speak as a Christian who values the interfaith spirit of our age and the age-old tradition of humanist learning.

While language is the means by which we convey meaning, I believe there is too little reflection on how the language of mathematics and the language of religion may share common characteristics. Stating that they are two alien types of language is too simple. Therefore, this book offers models for how the languages of science and religion complement each other. Science and religion live in a complex relationship (what in the last chapter I call Non-Symmetrical Magisteria). But they can both offer valid truths based on a common criterion of internal consistency and usefulness in the world.

I hope that this little book conveys to those curious about mathematics the rich world I have found in this discipline, and why I believe that seeing mathematics in new ways can increase our sense of the beauty of the world and our ability to find harmony between science and our faith traditions.

Mathematics *and* Religion

CHAPTER 1
Mathematics and Natural Sciences

SINCE THE RISE of modern science in the sixteenth century, mathematics has often been characterized as the language of nature. We often forget, however, that we have never stopped debating whether we can talk most accurately about the world by using only numbers or by also using physical models. Are the solar system and the movements in the night sky, for example, best understood by a series of numbers written on a sheet of paper, or when we view a wooden model of the planets and the solar system, so to speak, as the early scientists of the Renaissance did?

Although mathematics and natural science are closely bound together, they represent essentially two different kinds of language. Mathematics refers primarily to objects of the mind. Natural science refers to objects of our sense experience. In mathematics we use abstract formal signs (that is, the language of precise mental meaning and a language that we can manipulate mechanically). In contrast, natural science uses what we may call representational language that speaks of the physical objects which physics, chemistry, geology, and neuroscience study.

We can go deeper as well. At the heart of both mathematics and natural science lies the primary level of logic. Once we have logic, we are able to move on to mathematics and to natural science. At each of these levels, we perceive reality and then we use a type of language to express that perception.

FORMAL SIGNS IN LOGIC AND MATHEMATICS

What we perceive at the level of logic is correct reasoning, an inference that one thing naturally leads to another. We can test such logical inference in formal models of logic or mechanically, as in a computer. But many times we perceive something as logical simply by the power of intuition: it immediately seems to be so. These are logical intuitions. They intuit that something is following the rules of logic. For example, "It is impossible that something be true and false at the same time" is a logical principle that we intuit is always valid. We call this the principle of noncontradiction.

The logic we intuit can also be put into a formal language. As evidenced by the abstract signs often seen in logic or mathematics, formal language consists of a finite series of signs that follow rules of syntax. The signs have no definite meaning until they are related to each other by these rules, and then we can interpret these strings of signs as true or false. That language of logic sets the stage for the language of mathematics.

The way to understand the relationship of logic and mathematics is to say that while mathematics includes logic, it cannot be reduced to formal logic. Mathematics has something more, a kind of mathematical intuition and freedom based on logic. In fact, if we reduce mathematics to pure formal logic, we end up with paradoxes, which amount to contradictions. The great mathematical ambition of the German Gottlob Frege (d. 1925) and the Englishman Bertrand Russell (d. 1970), both of whom wanted to reduce mathematics to formal logic, illustrated this paradox. The result, however—which they conceded—is that such an effort ends in paradoxes. So again, logic and mathematics are different despite many similarities.

Like logic, however, mathematics also begins with intuitive perceptions. Mathematics begins as a purely intellectual, intuition-driven exercise. One of the first great mathematicians, Euclid, proposed many of these natural intuitions. For example, the first Euclidean postulate expresses the mathematical intuition that between any two points a straight line segment can always be drawn. In applying mathematics, we give these intuitions another name: mathematical axioms, which amount to beliefs that we presume to be true. (Hereafter, we use the terms "axiom" and "postulate" interchangeably since they have the same meaning.)

So mathematics is built of two parts, the axioms and the mathematical statements that seem logical. However, since axioms are basic intuitions, and they are the foundation of a particular mathematical system, axioms are not valid in all systems. What remains valid in all systems is the logic of mathematical propositions. As we see later in the book, this realization has created a variety—or pluralism—of mathematical systems. Yet in each one, certain logical propositions must always be valid. We can turn again to Euclid to illustrate this point. The axioms that Euclid began with ensured that his geometry was consistent and logical. However, not all forms of mathematics begin with Euclid's axiom. Basic arithmetic does not use those axioms, and thanks to modern revolutions in math, today we have non-Euclidean geometry, which uses axioms different from Euclid's.

As we can see, if axioms and logical principles are mixed in the wrong ways we end up with paradoxes, which means that we can deduce a proposition, but also its negation. It would seem that paradoxes would always be a bad thing, since they suggest that reality is not truly logical at all. However, the value of paradoxes is that they stimulate us to look more deeply for the logical connections in our intuitions and prove them in the language of logic or mathematics. While some paradoxes seem insurmountable, they also stimulate us to look beyond the use of the purely formal language of formal signs—used exclusively in logic and mathematics—to employ the representational language of empirical science and even the symbolic language of metaphysics (see Figure 1.1).

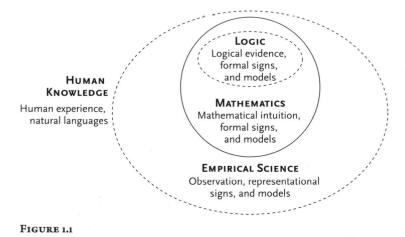

FIGURE 1.1

Representational Signs in Natural Science

The natural sciences begin with perceptions of the objects in the world, which is what separates them from the purely mental starting point of logic and mathematics, though natural science employs logic and mathematics as well. The empirical observations of natural science can be very sophisticated. Still, they are limited by the perceptions of the five senses. Once making their perceptions, scientists may certainly express them in natural languages, just as Copernicus spoke in Polish or German, but used Latin as academic language when he talked about his belief that the sun was stationary and the earth moved. Natural science ultimately seeks a high precision in its use of applied mathematics. Here mathematics becomes a privileged language; scientists understand each other and can conduct identical experiments despite their different national languages.

In practice, of course, the argument that natural science uses representational signs differently from the way logic or mathematics uses formal signs is a bit more subtle and complicated. The same signs can be used in either case. For instance, the signs E, m, and c in the equation $E = mc^2$ can be either formal or representational.

As formal signs they stand for elements of a mathematical structure, such as the system of real numbers (one of our more complex and inclusive sets of number systems). As representational signs, E, m, and c represent energy, mass, and speed of light. The difference between the two uses lies in the fact that, in the former, the semantics refer to mental objects (such as pure numbers) while, in the latter, the semantics refer to physical observations.

But we should emphasize again that logic and mathematics are at the core of natural science. Mathematics is indispensable in scientific research. The instruments that natural science uses to measure physical observations are designed based on mathematical theories. Moreover, logic and mathematics are not merely languages alone. They are the basic logical and mathematical intuitions that we cannot separate from our empirical sense experiences, and the bridge between those intuitions and our sense experiences has, in the history of science, been the building of scientific models.

FORMAL AND REPRESENTATIONAL MODELS

When we create models, we have structures that help us imagine how the world works. Models are mediators between perception and theories. In science, these models designate and describe the relations between the parts of a given domain of discourse and the procedures we can use to analyze the topic of research. The domain of discourse contains all the elements to consider in a given model.

Naturally, science builds formal models of logic and mathematics, and it also builds representational models that describe empirical observations (such as the wooden solar-system model of early astronomers). The first (formal) is real, but it is purely conceptual and does not have to necessarily match the reality "out there," for it only needs internal consistency. A representational model, however, must somehow match the empirical reality that an ordinary person can observe.

The models may be used together, depending on the problem that science is trying to solve. I mentioned earlier the model of non-Euclidean geometry, which essentially speaks of something we call curved space, as distinct from the normal flat space of geometry. So a model of non-Euclidean geometry can be created to talk about a reality that is not known to us; it is speculative, in this sense. But also, a model of non-Euclidean geometry can be a representational picture of the physical reality spoken of by Einstein's theory of relativity, which is a mathematical theory verified by observing the curvature of light in space.

For another illustration of how these models interact, we can turn to the story of Nicholaus Copernicus in the sixteenth century. In his day, the earth-centered model of Ptolemaic astronomy—essentially based on Aristotle's physical model—had dominated Europe for more than a thousand years because this representational model succeeded in explaining what astronomers saw in the skies year after year. However, Copernicus offered a mathematical model that explained the physical observations just as well—and more simply than Ptolemy's model of circular orbits and epicycles.

Let's reconsider Einstein's work. His theory of relativity was a purely mathematical model since he was not an astronomer (and, indeed, neither was Copernicus for the most part). Einstein built in his mind a model that tried to create a logical system to explain the universe on scales that were too large to measure physically. Einstein's mathematical model was tested in the empirical world in 1919 when the English astronomer Arthur Eddington journeyed to the Principe Island during a total eclipse to measure, by photography, whether curved space bent light as Einstein's model argued. In turn, the result was that Einstein's mathematics could explain the photographs (rough and questionable as they were). Today, our common scientific language refers to curved space and the four dimensions of space-time—representational models based on the formal mathematical concepts.

Another example of mathematical models rivaling representational models came about in the debate over the smallest scales of matter, as was seen in the difference of opinion between two of our greatest modern physicists, Niels Bohr and his student Werner Heisenberg. They both tried to address the problem of the uncertainty of the position and velocity of particles at the level of quantum physics. Bohr preferred a visual model of the solar-system atom, and to resolve quantum uncertainty, he ended up with a somewhat paradoxical image of the atom: he said that we can have two opposing yet complementary, visual models, one with the particle as a wave and one with the particle as a tiny corpuscle. On the other hand, Heisenberg preferred a single mathematical model of probability to explain how a real particle can exist without clear coordinates in the physical atom—a model of how it can be a wave and a particle at the same time.

The lesson here is that in the history of science, empirical observations have usually been interpreted in more than one way. We refine our knowledge by trying to reconcile a mathematical model with a physical model that we can observe. We are most satisfied when this harmonizing of math and observation works very well, but in the mysteries and complexities of the universe, there is no guarantee. In Figure 1.2, we see how we begin with both physical and mathematical perceptions, and then work our way up to models and languages, and then try to reconcile the languages.

Though we speak in Figure 1.2 about mathematical, empirical, and academic languages, we are also bound by our natural languages. We are all born into different cultures that have long tried to describe what we perceive in the world, whether that description is spoken or written in German, Hindi, or Chinese. As noted earlier, scientific language helps us transcend our local languages, although the transcendence can seemingly never be complete. Even in scientific culture, a pluralism of points of view exists—a variety of cultural languages. These many scientific communities have their own journals, congresses, and rituals. At times each scientific community seems to be, in effect, a different country.

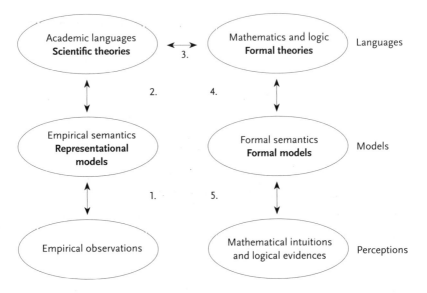

1. The empirical observations are explained through representational models of the reality. For example, Bohr's atom is a representational model of the interactions observed between the nucleus of the atom and the electrons. The representation of space-time in a four-dimensional space in the general theory of relativity is another example of a representational model of the empirical observations regarding the basic nature of matter.

2. The first explanations of the models of reality are expressed in an academic language. Academic languages are natural languages improved with some mathematical formalism. The use of mathematical formalisms in the explanation of the scientific models of reality cannot be separated from their explanation in a natural language.

3. The scientific theories can be explained by formal mathematical theories. For example, the laws of classical mechanics can explain the circular movement of the electrons around a nucleus in Bohr's atom, with the exception that only those orbits whose angular momentum is quantified are permitted. Another example of a formal theory that explains the models of the reality are the geometries of Hermann Minkowski through which the representational model of the theory of relativity is explained.

4. The formal mathematical theories are interpreted in formal models. For example, the formal theories of Minkowski are interpreted in a formal mathematical model of a non-Euclidean geometry.

5. The formal models of mathematics represent mathematical intuitions and logical evidences.

FIGURE 1.2

This situation reminds us of the difficulty in achieving pure objectivity in our perceptions, language, and model building. Even the purest formalism of logic cannot escape a degree of subjective interpretation. As we shall see next, even mathematicians disagree on what is absolutely logical.

FORMALISM AND OBJECTIVITY

Although we may affirm that logic and mathematics are the most objective knowledge, they are not totally objective or totally independent of the knowing subject. The view of what is logical and what is mathematical depends on the principles of logic that we accept. Some communities of mathematicians accept certain logical principles that other communities do not accept.

Even though the logical processes of deduction—with formal syntax and formal semantics—are objective and automatic routines that a machine can execute, several different possible views are available of what logic is. Accepting one or another view of logic can depend on personal tastes and preferences for what counts as valid logic. One can assume different views of logic, but not at the same time if they are not consistent. Once a view is assumed, one must maintain consistency.

Surprising as it may seem, not all logicians accept the famous principle called the excluded middle, for example. This principle states that "all propositions are true or false." Classical logic is strongly rooted in the principle of the excluded middle. But other schools of thought in mathematics, such as the constructivist or intuitionist schools, do not accept this as an absolute premise. This debate over the excluded middle is a disagreement about the existence of what we call mathematical objects. The classical view is that to prove the existence of a mathematical object it is enough to derive a contradiction from the assumption of its nonexistence. According to the contrary (constructivist) view, one must find (or construct) any mathematical object in order to prove its existence.

The classical approach to mathematics strongly defends the principle of the excluded middle by reducing the contrary view to a position of absurdity, a common method of proof we call "reduction to the absurd" (or reductio ad absurdum). That is the idea behind the following scheme:

A or not A (Logical principle of excluded middle)

A implies B (Premise)

not B (Premise)

∴ not A (Conclusion)

In the case that A is true, B should be true, because A implies B. But B's existence with the premise not B would be absurd. Therefore if the logical principle of excluded middle is accepted, not A should be our conclusion.

In another example, using reduction to the absurd it can be proved that $\sqrt{2}$ is a number with infinite decimals. In order to prove this, it is sufficient to see that if $\sqrt{2}$ is a number with a finite number of decimals (A), then $\sqrt{2}$ could be written as the quotient of two whole numbers (B). Once we admit this fact (A implies B), it is enough to prove that $\sqrt{2}$ cannot be written as the quotient of two whole numbers (not B). Then, by the principle of excluded middle, we prove that $\sqrt{2}$ is a number with infinite decimals (not A).

This proof is valid in classical mathematics. Classical mathematicians admit the existence of sets with an infinite number of elements, whereas constructivist mathematicians do not admit the existence of such sets. What is evident for some is not so for others. Classical mathematicians believe that the existence of a mathematical object is proved if a contradiction is derived from its nonexistence, by reduction to the absurd. Constructivist mathematicians simply disagree. The members of the two groups chose their position based simply on personal preference, a valid enough reason for why people join one group or another, but a reason that is independent of the logical evidence.

The different views of mathematics—the classical and the constructivist—are based on different views of logic. When the Dutch mathematician L. E. J. Brouwer (d. 1966) stated that the principle of the excluded middle could not be applied in all cases, he was arguing for a new kind of logic. His statement was not supported by a logical deduction. Still, Brouwer persuasively justified his statement based on the meaning that the totality of mathematical activity had for him.

Another way to look at this fact of two or more systems of logic is to recognize that they all can be formalized into a rule-based system that can operate, for example, on a computer program. We can program a computer so that it executes classical deductions or constructivist deductions, or another type of deduction entirely, depending on the circumstances. However, when we speak of mathematicians as believers in one or another type of logic, then we are talking about personal preferences—and these typically appear as schools of thought.

This pattern of individual options and different groups of scientists illustrates that even when we speak of a universal activity such as logic or mathematics, these disciplines are not purely formal and mechanical. Logic does not easily transcend our normal human subjectivities. The student of logic can choose one logical principle or another, but there is no formal, logical argumentation—written in the heavens or on stone tablets, so to speak—that can help him to decide which one is best. The decision, in effect, is not a purely logical one.

That there are several views of logic and that these views depend on communities and their preferences can appear surprising. Indeed, the plurality of logics seems to contradict our impression that mathematical propositions and deductions have a high standard of certainty. Can logic and mathematics truly be universal languages?

PUBLIC LANGUAGE

Despite this pluralism, logic and mathematics continue to be our most objective—and therefore privileged—instruments for public communication of knowledge. For a start, mathematical signs have the same meaning for everyone, whatever the circumstance. Within each mathematical system, all the elements are equal as causes of their relationships. Mathematical language has no room for the description of a final—that is, ultimate and directing—cause. Mathematics does not offer a First Cause or a Final Destination in the way it explains nature. This removes from mathematics the temptation to include the biases or agendas of human beliefs, making mathematics a useful public language.

Natural science also has its public language, thanks in great part to its relationship to logic and mathematics. Mathematics allows us to have a public language about physical, chemical, and biological realities. We can talk publicly about Isaac Newton's Second Law because its principle that force is proportional to mass and acceleration can be stated formally as F = ma. This principle can be explained in Chinese, Russian, or Swahili. The Darwinian theory of evolution by variation and natural selection has also entered our public language because scientists have observed and quantified many of these processes, and some of them have been written in mathematical language.

Every scientific observation cannot end up with a purely logical and mathematical presentation, of course. The world remains vast and paradoxical, not surrendering everywhere to human reason. So, in the case of the quantum physics of the atom, Niels Bohr has argued that we must simply use two kinds of mathematical objects to explain the apparent duality of particles as waves and corpuscles. Bohr calls this the principle of complementarity, presenting the case that we can rationally use two models to explain a single physical object in the universe. A particle can have two possible behaviors, and each of them can be explained by a mathematically consistent description. In physics, Bohr's principle of complementarity is famous for what it says about the limits of human perception. In modern culture, complementarity has also become the slogan of contemporary movements and beliefs that, often being antiscientific and postmodern, emphasize human creativity, paradox, and mysticism.

The complementarity spoken of in physics and in popular culture is not exactly the same as complementarity between the languages of logic, science, and metaphysics. Still, there is a similarity. The idea that the signs of logic and math positively complement the symbols of metaphysics has a parallel in physics, for in both cases, the two explanations are offered for one reality. Furthermore, reality transcends each of the explanations in both cases.

We mentioned earlier that the structure of mathematical language does not allow for the insertion of ultimate causes or ultimate outcomes. To find those ultimate causes we need the language of philosophy, metaphysics, or religion—we need a symbolic language. This language is nonmathematical and yet is able to create a self-consistent system that can speak to individuals and communities. In symbolic languages, we accept a plurality of ultimate meanings, causes, and reasons. We accept their plurality even though we live in the same mathematical and physical world.

Typically, we turn to such metaphysical language because neither formal logic and mathematics, nor empirical science and its models, can answer every question or solve every paradox.

CHAPTER 2
Metaphysical Language

IF WE WANT to keep our lives as simple as possible, mathematics and natural science offer a great advantage. Neither of them asks ultimate questions. Metaphysics, in contrast, is about ultimate things. For this reason, metaphysical questions may complicate our lives, but they also help us to resolve some of the deepest mysteries. Mathematics and natural science accept reality as a given fact; the questioning ends there. However, metaphysics asks why our minds are able, in fact, to understand the physical world and mathematics. Indeed, metaphysics asks why things exist at all. Why is there something rather than nothing?

Offering a more global and radical way of asking questions about reality, metaphysics points to the possible existence of an ultimate principle that justifies the existence of things in general. For most people, the metaphysical questions are unavoidable. But we also know from history that many kinds of perceptions of ultimate things exist, and therefore many kinds of answers to metaphysical questions. These questions and answers also form distinct communities, such as cultures or religions, just as we have seen in mathematics.

For example, one group may look at the metaphysical evidence in life and arrive at a basic intuition that the world exists on its own and for its own reasons. We may call this a pantheistic metaphysical view. Another kind of ultimate view may be called agnostic. It argues that we cannot know ultimate principles, so whether they exist or not is irrelevant. Third, a theist may believe that the universe exists because God exists and this Creator has made and maintains the universe. Finally, another metaphysical school may say that the ultimate principle in the universe is mathematics itself, as we shall see with schools such as the Pythagoreans and certain Platonists.

Metaphysics, of course, uses a different language from logic, mathematics, and natural science. This is the language of symbols that stand for ultimate realities or ultimate types of relationships. These symbolic words can range from God and the cosmos or universe to words found in mathematics, if that is deemed the highest reality. Whatever the language/symbol in metaphysics, it exceeds the meaning of the signs of mathematics and of the natural sciences. The word "number" is precise and definable in mathematics, but when used by the ancient Pythagoreans, for example, "number" not only refers to a mathematical object but also to the ultimate foundations of the world.

A scientific term is fairly objective when it speaks of a measurable object. A metaphysical term/symbol must be approached differently, however. A metaphysical or religious symbol is understandable only within a history, a tradition, and a community which uses that symbol. The symbol and its context provide its coherence. That context is empirical, for it is made up of history and tradition. What is more, metaphysical symbols can refer even to "Nature." But this is different from the physical measurements of natural science. Metaphysical symbols are, again, mostly determined by the community that uses them to speak of realities beyond what is empirical.

At one point in my life, my interest in both mathematics and theology was deepened by a particular type of metaphysical question. I had completed my advanced studies in mathematics and was pursuing theology when my professor stated the following idea: "The world is totally related to God, being totally different of him." On one hand, this proposition is about a kind of relation, which is what mathematics is all about. This relationship (regarding God) could be written in mathematical notations, just as we can convey the idea of 2 + 2 = 4 as a relationship of factors that not only add up but also include each other and are beyond each other. But in a theological statement such as "The world is totally related to God, being totally different of him," more than just a formal (logical or mathematical) relationship is being conveyed.

This metaphysical, and therefore symbol-based, statement about God and the universe is possible because it is made in the context of a tradition and community—in this case, the Christian tradition. The symbols have had a resonance of personal meaning as well as a formal kind of logic. In metaphysical terms, the symbols speak of God's transcendence. They suggest that while the world depends on a Creator, the Creator does not depend on the world. For me, the idea had a transformative effect. Such symbolic meanings are about more than the mere objects related. The meanings can transform us and our communities.

Scientific Meaning
and Metaphysical Meaning

In some ways, a scientific hypothesis also is a symbolic picture of how the world might be. But in science, we then try to test and verify that hypothesis. Under these tests, some hypotheses fail and others survive to be tested further. We test the truth of metaphysical and religious formulations differently. They are not subject to empirical testing as we would test scientific principles in a laboratory. Instead, metaphysical and religious formulations seem true to us when they offer an intuitive veracity and coherence in the context of the personal values that a group of people share. Those people outside the context naturally have a very hard time judging the intuitive truth and coherence of a metaphysical viewpoint. As years of interfaith dialogue have shown us, when a Christian states that Jesus is the truth and fullness of God, the Hindu, with another kind of tradition and community, cannot easily comprehend the statement.

Another difference between natural science and metaphysics is that in science, in principle at least, it is said that a hypothesis and theory are never absolute, but only the best approximation so far known in science. In metaphysical perception and language, however, the goal is to arrive at an absolute, a framework that can serve as a permanent vision of reality and the values that flow for it.

Because scientific hypotheses do not try to arrive at the finality known in religion, scientists can hold many metaphysical views. They can be agnostic, atheistic, or theistic in their fundamental beliefs. In other words, a scientist might easily understand the idea that God cannot be tested or proved because God, by definition, is not a mundane object. Religion offers a wider meaning to experience and history, even a timeless meaning.

When metaphysical symbols become articulated, they offer what we call myths, or grand narratives common to all religions and stories of the origins of the universe, or of how the human predicament came to be. (Myth, in this sense, does not mean an intentional falsehood, but rather a traditional story that tries to meaningfully unite the facts and mysteries of life.) A community shares these symbolic narrations and passes them on to subsequent generations. The heart of the myths, however, is a set of values that endure even when some of the religious language or narrative changes by its interaction with newly discovered facts about the world. For example, in Christianity, the stories of the human fall into original sin and the Resurrection of Jesus both convey the human struggle with evil and hope for salvation, even though these myths can be told in different ways in Christian literature and theology based on our experiences in the world.

To be sure, science has its myths and its narratives as well. One of the most famous is the idea of progress, that science will always and everywhere find true answers and make the right decisions. At the heart of this scientific myth is the belief that everything changes, usually in an upward progression. But again, this symbolic kind of vision is hard to test in the laboratory. We seem to accept this myth as true, though, even though alternative views say that some things may not change, or that progress may be relative and elusive.

As increasing knowledge and communication in our world challenge many of our historical myths—religious and scientific—the myths that serve us best try to reconcile themselves to the empirical findings of science. Our most helpful myths cannot contradict the scientific data. Religion can accept the knowledge of nature that science provides. In turn, science hopefully can recognize its own necessary openness and its own myths and make room for a dialogue with the metaphysical languages that guide our human communities. This is more of a meeting of value systems than a battle over the facts of the world. The meeting is not a contest of pure logic, but a discussion of ethics, behavior, and history as well.

Nevertheless, there are fascinating times in our human experience when scientific and metaphysical concepts have overlapped in a very compelling way. As the rest of this chapter shows, two of those concepts are infinity and what has come to be called the ontological argument for the existence of an absolute—that is, for the existence of a supreme being.

The Infinite and the Ontological Argument

From the beginning of human knowledge, we have wondered how to conceive of the idea of infinity, something that goes on forever. We also wonder how this concept helps us in our lives, especially in mathematics and science. At the same time, we have pondered what it means for something to be the highest, greatest, or absolute in our world. As we will see, this concept was often expressed in the idea of God as "the greatest that exists" or "the greatest that can be thought." Remarkably, both of these ideas—infinity and the greatest realm of existence—have been parallel questions in math, science, and metaphysics. We begin with the concept of infinity.

Infinity: Mathematical, Empirical, and Metaphysical

For mathematicians, the infinite can be a sign with two distinct meanings: actual infinite and potential infinite. The potential infinite is a succession of mathematical objects that can be as big as we wish, because it does not end. For example, starting from a mathematical object that we designate as 0, we can construct its successor and designate it s(0). Then we can define the successor to this object and designate it s(s(0)), and then to s(s(s(0))), and so on. Thus, we can define the set N = {0, s(0), s(s(0)), ...}, which has as many elements as we wish. Then we can say that the number of elements of N is *potentially infinite*.

In contrast, we can speak of the *actual infinite* when we refer to all the numbers of N as a whole. When we consider N as a whole, we observe all the possible elements of N globally. (In the potential infinite, we consider N as a potentially infinite set observed not as a whole, but as a process of constructing its elements.) This distinction between potential and actual became important in the work of German mathematician Georg Cantor (d. 1918).

Cantor proved that the consideration of N as a "whole" (the actual infinite) leads us to consider other infinites that are "greater" than N, which usually are called transfinites. Some mathematicians could not accept this state of things. Now called constructivists, such mathematicians could admit only the existence of those mathematical objects that can be constructed based on finite symbols and functions. Hence, constructivists admit the existence of only the potential infinite in mathematical practice. They argue that the actual infinite cannot be constructed by finite means in a finite process.

As we noted earlier, such disputes allow mathematicians to choose the system they prefer. If we choose classical mathematics, then we accept the actual infinite. A constructivist in mathematics accepts only the potential infinite. Both camps have a series of arguments to justify their choices in applying a mathematical approach to infinity.

In natural science, the concern is to arrive at an empirical idea of infinity that is useful for scientific hypotheses and research. Natural science draws on the mathematical idea(s) of the infinite, but tries to apply it in representational language that talks about the physical world. To begin, we must distinguish between the representational sign and the physical object represented. The infiniteness of time and space as a physical property of the world, for example, is often confused with mathematical infiniteness. To clarify, let's look at some of the physical descriptions of infinity.

Natural science has always struggled with the problem of trying to measure something that is infinitely small or infinitely large. Some physical cosmologies have seen time and space, for example, as infinitely large, and thus unmeasurable as an empirical object. However, current physics tends to think of the physical universe as being a limited object. In the big bang theory, for example, time originated at the original explosion, and time will end after a certain period of expansion of the physical universe. Indeed, once the expansion is over, a big crunch may ensue: a time period when the universe collapses back to its original single point under the forces of gravity. The big bang theory also suggests that space is finite as well. Real space would end where the universe ends. At the smallest levels of physics, quantum theory similarly argues that there is no infinity to matter: quantum physics says it is not possible to think of infinitely small measurements. As we can see, empirical science offers a few options concerning the infinity of time and space. Each is based on a different vision of the physical world.

The metaphysical view of infinity might have been the first one that human beings grappled with, and always in symbolic language. In metaphysics, the symbol of the infinite stands for what is absolute, whereas the symbol for the finite means the relative—or what is dependent on or contrary to the absolute. Typically, metaphysics views human existence as a finite reality in relationship to an absolute reality. This also puts limits on human knowledge, including logic, mathematics, and empirical science.

Despite this contingent nature of the human being, a kind of absolute consistency still controls how our minds approach logic, mathematics, and science. Our thoughts cannot cease to be consistent. Our minds cannot affirm A and not A at the same time. We cannot state that reality is logically contradictory and continue to think logically. Science is not possible if our thought ceases to be consistent.

The acceptance of the absoluteness of the consistency of thought is previous to the mere act of thinking. Yes, our thoughts harbor many inconsistencies. We often think of something in one way and then think the contrary five minutes later. However, we know that we changed our thinking in the past five minutes because we have an underlying consistency in the mind. In this argument that the mind has a core element that is absolute we also find the reason that human beings can communicate: they share this consistency in thought.

Now we arrive at one of the great metaphysical questions: does that mind have this consistency because of an absolute in the universe, or simply because of the powers of the independent human mind based, for example, on the evolution of the physical brain? If we choose a higher absolute in the universe, then we opt for the metaphysical infinite, typically given the name God. If we say the apparent consistency of the mind arises from its own finite limitation, then we do not need an absolute to explain our world. This debate, or puzzle, is the basis for the famous ontological argument for the existence of God, a surpassing and infinite absolute.

The Ontological Argument

As I hope to show, the ontological argument is relevant to our discussion about mathematics and language because the argument lies between metaphysics and mathematics. The historical ontological argument has typically employed some of the logical tools that underlie mathematical theory.

Before we go further, we should define our term. "Ontology"

means existence. In the eighteenth century, the logical argument for God's existence as a necessary absolute entered Western philosophical writings under the name of the ontological argument, or the ontological proof of God. While some Greco-Roman writings contain a simple form of the argument, it was formulated most precisely by Islamic and Christian thinkers in the Middle Ages.

Any type of reasoning which concludes that God's existence is logical includes our metaphysical intuitions, just as logic and mathematics begin with intuitions. In philosophy, moreover, the ontological argument is unique for how it tries to show that the existence of God is a reasonable conclusion. Its core is the intellectual intuition that it is logically necessary for there to be a reality that is not limited by finiteness. Interestingly, this intuition may exist apart from religion, for as we will see, some mathematicians argue for an infinite absolute. Obviously, though, the ontological argument is a mainstay for religious belief and theology.

But even in religion, different schools of thought have found the ontological argument either helpful or not. For example, the argument—first elaborated by the Muslim philosopher Avicenna (or Ibn Sīnā) of Persia (d. 1037) and the Catholic theologian Anselm of Canterbury (d. 1109)—was also accepted in one form or another by René Descartes, Baruch Spinoza, Gottfried Leibniz, and later the mathematician Kurt Gödel. However, it was also rejected by so great a Catholic philosopher as Thomas Aquinas, and then again by Immanuel Kant, both of whom nevertheless did not deny the existence of a supreme being.

The work of Avicenna and Anselm offers a simple basis to follow the historical legacy of the argument. In *The Book of Healing*, Avicenna made the important distinction between the essence (*mahiat*) and the existence (*wujud*) of beings. He then postulated the necessity of a highest being whose existence is necessary (*Wajib al-Wujud*). Anselm is perhaps the most famous Christian thinker to build upon a similar kind of logic. He said that if a being "than which nothing greater can be conceived" did not exist in reality as well as in the mind, then it could not be the greatest being. Therefore, God must exist, for otherwise logic itself is rendered absurd. As Anselm said in his small work *Proslogion*, speaking in the devotional manner of all his writings, "We believe that You are something than which nothing greater can be thought. . . . And surely that-than-which-nothing-greater-can-be-thought cannot exist in the mind alone. For if it exists solely in the mind, it can be thought to exist in reality also, which is greater." Descartes, Spinoza, and Leibniz all built upon this basic kind of logic; they tended also to argue that since a necessary quality of absolute perfection is to have existence, then God must have existence. In the twentieth century, Gödel revived the validity of Anselm's logic by turning it into a purely formalized argument that was valid in modal logic, a system of logic that looks at what is possible, what is impossible, and what is necessary.

This relationship of the ontological argument to logic and mathematics remains a fascinating topic for us today, including for some mathematicians. Most important, it shows us the difference between the use of mathematical and logical signs and the use of metaphysical and religious symbols. The ontological argument cannot be made, so to speak, in the formal signs of mathematics alone, for it also requires the addition of metaphysical symbols.

The main criticism of the ontological argument is that it tries to prove with symbolic language (the greatest that can be thought) what requires a mathematical or physical reality to prove. For example, the critic would also say that while the ontological argument uses formal logic, it does not prove the existence of God in the same way that a mathematical proof verifies the existence of a mathematical object, such as the fact that, given a prime number, there is always a greater prime number. In verifying a mathematical object, such a proof begins with a mathematical intuition. In the case of a proof for the necessity of God's existence as an absolute, the formal logic is based on a distinct metaphysical intuition—the presumption, that is, that such a universe with an absolute seems appropriate. For the critic, there is no evidence for this beginning intuition.

As noted earlier, the two well-known opponents of the ontological argument were Aquinas and Kant. Both of them argued that from the basic intuition about existence, it is not possible to argue that something real must necessarily exist as well. Aquinas and Kant, in other words, questioned whether we can leap from what is logical to what is real. They believed the flaw of the argument is that it confuses the formal semantics of mental objects with the real semantics of objects in the real world.

For our purposes, the key question is one of defining existence—something that mathematics, empirical science, and metaphysics all try to do as a beginning premise, very much as an intuition of what is logical and appropriate. How does mathematics define existence? When we state that a solution exists to the equation $x^2 = 4$, we are saying that mathematical objects exist. In this case, the objects are the numbers 2 and -2, which are the solution to this equation. The numbers 2 and -2 can be formalized and represented as formal sets.

Empirical science has another kind of definition of existence. A compelling example of how science affirms existence came on August 24, 2006, when the International Astronomical Union (IAU) ruled that Pluto can no longer be considered a "planet" and the number of planets in our solar system was thus reduced from nine to eight. This kind of definition relates to the existence of empirical objects: the planets are physical places to which our spaceships could travel. But they are also objects that we place in categories of existence, such as planet or nonplanet (as is the case now with Pluto).

Metaphysics has yet another definition of existence: the global question of how contingent, or finite, beings have a relationship to being itself—that is, a necessary supreme being. Spinoza posed this question in his famous philosophical work, *The Ethics*. He said that God necessarily exists because God's essence is to exist. Spinoza was making a metaphysical statement about the existence and essence of God. For Spinoza the existence of God coincides with the essence of God, unlike the existence of all the other beings, which are distinct from their essence. This identification of essence and existence in God and their separation in other beings is valid insofar as it is confirmed by a metaphysical intuition or experience.

We refer to Spinoza's metaphysics as pantheism, since he closely ties the physical world to God's existence. Thinkers such as Anselm and Descartes, however, came down on the side of traditional theism by saying that while God's essence allows other beings to exist, God's existence is separate from that of all other beings. Despite these differences, all of the ontological arguments agree on the metaphysical intuition that there must be a being without limits. To quote Anselm again, this is a being such that we cannot think of any greater being. Based on this premise, all of the ontological arguments use valid formal logic to show that the existence of God is reasonable.

Formalization of Ontological Argument in Predicate Logic

Now let's look at how such an argument, which might seem purely religious, can actually be made using the signs of logic and mathematics. For readers who are unfamiliar with the way logic is written as a string of signs, it will be enough to see that verbal statements about the world can be translated into these strings of notation. By looking at this use of signs closely, we can better understand the difference between formal signs and metaphysical symbols.

We start with a simplified formalization of the ontological argument (OA), which is based on a metaphysical sentence about the existence of something without limits.

OA:

(1)   God is the greatest being we can think of.

(2)   If God does not exist necessarily in reality, then we can think of an existing being greater than God.

∴   Therefore God exists in reality.

Taking the propositions (1) and (2) as premises, we can formalize the argumentation OA in classical predicate logic. (Readers who are interested can find a brief presentation of the syntax of first-order predicate logic in Appendix 3.) To do so, we first define the predicates GR(x,y) and E(x) as follows:

GR(x,y)   :↔   We can think that x is greater than y.

E(x)   :↔   x exists in reality

We can also use the letter g in order to formally designate God:

g   :↔   God

Using these signs, we formalize OA:

(1)   ¬∃ x GR(x,g)

(2)   ¬E(g) → ∃ x GR(x,g)

∴ E(g)

where we can read the premises (1) and (2) and the conclusion ∴ as follows:

¬∃ x GR(x,g)

There does not exist an element of our domain of discourse x such that we can think that x is greater than God.

(2)   ¬ E(g) → ∃ x GR(x,g)

If God does not exist in reality, then there is an element x of our domain of discourse such that we can think that x is greater than God.

∴   E(g)

God exists in reality.

In the formalization of the ontological argument, we have used two different signs, the logical sign ∃ and the predicate E, in order to denote the real existence of the same element x. The use of the logical sign ∃ in formula (2) indicates the existence of an arbitrary element x of our domain of the discourse to which we attribute the predicate GR(x,g). The use of the predicate E in the subformula ¬E(g) indicates the real nonexistence of g.

It is easy to verify that, by applying the rules of deduction of classical logic to the formal propositions (1) and (2), the conclusion (∴) follows.

In fact, classical logic admits the principle of *tertio excluso* whereby, given a predicate E and an object g, or E(g) or ¬E(g), one of the two is true. If ¬E(g) is true, by the conditional proposition (2) we would obtain that ∃ × GR(x,g) is true. This would make the principle of noncontradiction false as both ∃ × GR(x,g) by (2) and ¬∃ × GR(x,g) by (1) are true. Therefore, we conclude that if (1) and (2) are true, E(g) must be true.

The formalization of the ontological argument shows that we can construct a formally correct ontological argument and conclude that God exists, in the case that the two premises (1) and (2) are true. However, the problem stated by Aquinas and by Kant still is not addressed directly in the above solution. Aquinas and Kant argue that we cannot confuse existence as thought with existence in reality. They would argue that in premise (2) we are confusing the logical sign ∃, which refers to formal existence, with the predicate E, which refers to necessary real existence. That is to say, we are confusing the real world with the mental world.

How does the evidence that was so persuasive to Avicenna, Anselm, and Descartes in the ontological argument relate to the kind of evidence demanded by empirical science? Obviously, empirical science required confirmation by the physical senses. One good example is the geometry of the German mathematician Hermann Minkowski, who offered propositions about Einstein's theory of relativity. The Minkowski propositions must be tested against physical observation and measurement. A metaphysical proposition is different: it must be confirmed by the intuition that there must be such a thing as unlimited existence.

Supporters of the ontological argument as valid, therefore, would accept the formal proposition $\neg E(g) \rightarrow \exists \, x \, GR(x,g)$ as an expression of a metaphysical perception, intuition, or evidence. Again, the metaphysical intuition is not apprehended like the empirical object of a planet or the bending of light. It is also not like the purely formal perception behind the existence of a mathematical object. Metaphysical proof has its own way of being thought and felt. The use of the logical sign $\exists$ in formula (2) indicates the existence of an arbitrary element x of our domain of the discourse to which we attribute the predicate $GR(x,g)$. This element x can exist only in our mind. It is a metaphysical option to give it real existence. The use of the predicate E in the subformula $\neg E(g)$ indicates the real existence of g.

Anselm and other defenders of the ontological argument reasoned correctly from a formal point of view. Their conclusion is valid in the case that the following metaphysical premise is valid: "If God does not exist in reality, then there is an element x of our domain of discourse such that we can think that x is greater than God." But although that premise can be accepted as metaphysical evidence, it is not necessarily true.

We can also look at this from the viewpoint of predicates. For example, $GR(x,g)$ is a predicate with two arguments that we can interpret in a model of reality as saying: the element x of the domain of discourse is greater than the element g. On the other hand, we could say that $E(x)$ is a predicate with an argument that we can interpret as saying: the element x of the domain of discourse exists in reality. (Readers can find more details on the semantics of first-order predicate logic in Appendix 4.) The fact that the element g does not have the property E (of existing in reality) does not imply that x is greater than g. In other words, $\neg E(g)$ can be true and $GR(x,g)$ false.

These kinds of formal arguments can become more and more sophisticated, but they invariably move us further away from reality. In summary, the fact that we can think of an object that is bigger than God does not mean that such an object really exists. It may be an object simply created by the mind—a figment of the imagination. Once again we see the importance of choosing a beginning premise for our logical systems. Different communities choose different premises, and therefore some agree with the ontological proof, and others do not.

Abrahamic Religions and Metaphysics

Our discussion of metaphysics suggests that it has a natural fit with the monotheistic religions, which we also call the Abrahamic faiths, since they all trace themselves to the early monotheism of Abraham: Judaism, Christianity, and Islam. All of these traditions have relied upon tools of logic and metaphysics to show the reasonableness of belief in the existence of the absolute. However, they also go a step further than relying on a kind of logical argument based on the intuition that an absolute must exist. They also teach the need of having faith in a transcendent God that is beyond logic and mundane evidence.

This faith requires the belief that God, acting independent of the world, has revealed his existence to the world, an existence that the human mind cannot entirely apprehend. A kind of revelation given to the world must also be at play, according to this Abrahamic tradition. Often this metaphysical perception of God leads individuals to feel a personal mission they must undertake to the world. Many such testimonies are present in history. We can think, for example, of the prayers of the Englishman John Henry Newman (d. 1890), who explored both the Anglican and Roman Catholic traditions. "God has created me to do Him some definite service," wrote Newman, who became a Catholic cardinal and university educator. "He has committed some work to me which He has not committed to

another." In my own specific tradition, the founder of the Jesu-
its, Ignatius of Loyola (d. 1556), also offered an interpretation of
human purposes in the world in the opening section of his *Spiritual
Exercises*, titled "Principle and Foundation":

> The human person is created to praise, reverence and
> serve God Our Lord, and by so doing to save his or her
> soul. The other things on the face of the earth are cre-
> ated for human beings in order to help them pursue the
> end for which they were created. It follows from this that
> one must use created things in so far as they help towards
> one's end, and free oneself from them in so far as they
> are obstacles to one's end. To do this we need to make
> ourselves indifferent to all created things, provided the
> matter is subject to our free choice and there is no pro-
> hibition.

The language of such testimonies cannot be mathematical or scientific. It must be a language of theological symbols. Those symbols, in turn, are moral and ethical. They are more than a narrative,
for they also speak of the meaning of human action in the world
and a daily relationship of the human being to the absolute. This
language—often the passionate language of faith and belief—is
about attitudes and relationships, about such experiences as trust,
mercy, and forgiveness.

SUMMARIZING THE THREE LANGUAGES

At this point in our comparison of the languages of logic/math-
ematics, empirical science, and metaphysics, it could easily seem
that we are looking at three entirely different worlds. The gap may
seem especially large between metaphysical symbols and the signs
of math and science. However, the three languages can also be seen

as complementary because, by applying them all, we speak to the full reality of the human experience. None of these languages can be excluded. As we saw earlier, while logic may not need religion, religion does need to harmonize itself with logic and science.

The key difference between these languages relates to the use of signs and symbols. In our day-to-day language, we don't really see any difference between the words "sign" and "symbol," and indeed the distinction is of more interest to academics. But the distinction is helpful to everyone if we want to understand the interaction of our three kinds of languages: formal, representational, and metaphysical.

In our use of logic, formal signs make it possible to write propositions that express evidence. For example, the set of formal signs $\{\neg, A, \wedge, (,)\}$ makes it possible to write the proposition $\neg(A \wedge \neg A)$. This is the basic principle of noncontradiction. Furthermore, the formal signs of arithmetic allow us to express mathematical intuitions regarding numbers and certain numerical operations. For example, the set of formal signs $\{2, 4, +, =\}$ makes it possible to write the arithmetical proposition $2 + 2 = 4$.

Our second kind of language uses the representational signs of physics and the empirical sciences. This language points to observable realities such as the mass, speed, or energy of physical bodies. The set of representational signs $\{E, m, c\}$, gives us the ability to talk about energy (E), mass (m), and the speed of light (c) in the useful equation $E = mc^2$, which expresses a property of the physical theory of relativity.

Finally, because our minds are able to logically ask ultimate questions, we need a metaphysical language to speak of what mathematics and science cannot touch. These metaphysical questions and answers need a language of meaningful symbols that are neither simply formal nor empirical. The metaphysical meaning of symbols expresses personal evidence about the ultimate principles of reality. The people who specialize in logic, mathematics, and science arc all members of communities, and these groupings influence, even bias, the assumptions made in using these generally objective types of languages. This social context is even more important in metaphysical language, however, for only by having a distinct community—a history, tradition, and core values—can this language retain its powerful meaning.

CHAPTER 3
Origins of Mathematics

ABOUT TWENTY-FIVE centuries ago, humanity began an intellectual ascent that involved the construction and accumulation of the knowledge composing today's systems of mathematics. The summit of this accumulation cannot yet be seen, but we can look back and see the route we have walked.

In the broadest view, this story is about the development of human rationality. This rationality perhaps has its high-water mark in logic and mathematics, but these pursuits are not the only realms of human thought that have used the powers of reason. As we saw in the previous chapter, the human ability to ask metaphysical questions has evolved along with our ability to develop mathematical systems.

For all of the prowess of mathematics, we have encountered its limits. Mathematics is not absolute, and it is shaped by the choices of human beings and by the cultures in which it has developed. Although logic and mathematics are the most formal and objective of all forms of knowledge, mathematics is not completely formal and objective, because it also depends on personal and community preferences and choices. Mathematics is embedded in cultural meaning; as it developed in history, mathematics also took on a symbolic meaning. In other words, their discoveries often led mathematicians to metaphysical outlooks.

In this chapter's brief history of mathematics, we consider some of these discoveries. But first, let's look at a few illustrations of how mathematics can create communities of symbolic meaning even as it tries to be an objective language that transcends narrow groups of people.

For example, the mathematical insights of the Greek thinker Pythagoras also led to the formation of a religious Pythagorean school that attributed a mystical meaning to numbers: numbers were the creative principle of all reality.

Later in Greek thought, Plato asserted that certain numbers were among the eternal Ideas or Forms that lay behind the ever-changing appearances of the world. The concepts of Platonism influenced Western thought through the Middle Ages and even down to the present. Even in more modern times, mathematics has not been able to detach itself from its communities. For example, the English mathematicians who followed Isaac Newton adopted his form of notation for infinitesimal calculus, while in continental Europe, mathematicians used the calculus of Newton's rival, the German thinker Gottfried Leibniz.

All of these examples show how communities formed not only around mathematical discoveries, but also around symbolic and personal meanings that they shared in common. In the evolution of mathematics across history, it is hard to say that these systems and communities have simply built themselves one on top of another, as if constructing a house of historical mathematics. Nor is it wise to say that different math systems have simply been added on like new elements to a grand computer program. Instead, the relationships of the plural systems of mathematics are complex and interacting. Today, the many separate mathematical communities share a web of connections. They are often grouped as subcommunities around the mathematical disciplines of logic, algebra, analysis, geometry, or topology.

To look back at the history of mathematics, we must adopt a dynamic image of its evolution across different epochs. We cannot separate its development from the development of human culture, which includes art, literature, philosophy, and theology. With this thought in mind, we nevertheless look at mathematics as the backbone of the development of human rationality. Mathematics has influenced scientific thought as it also has given us a variety of rational systems for understanding our world.

In the following four chapters, I chart this evolution across four epochs, each with a major characteristic in the development of mathematics (see Figure 3.1). This fourfold scheme begins with the era of primitive mathematics, when its rudiments were first discovered in several cultures, especially in Africa and Asia. Next is the period of the early Greeks, who began to collect a system of mathematics that was expanded upon through the Middle Ages. The third period comes with the scientific revolution of the seventeenth century, when Galileo, among others, declared that mathematics was the language of nature. Finally, in the twentieth century, the language of mathematics was formalized, giving us the tools we use today to analyze entire mathematical systems and the powers behind our information revolution.

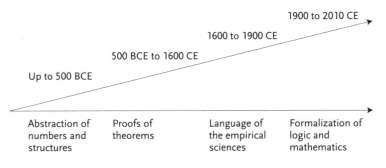

Four Epochs of the Development of Mathematics

FIGURE 3.1

PRIMITIVE MATHEMATICS

Sometime in the distant past (perhaps around 5000 BCE), our ancestors put the first and most basic mathematical capacities into practice. We can summarize this as a group of six accomplishments:

1. Providing language for numerical intuitions
2. Representing numerical amounts and geometric structures
3. Representing relations between numerical amounts
4. Creating calculations based on numerical relations
5. Quantifying intuitive-empirical knowledge of some
   mathematical theorems
6. Developing the concept of the infinite

These accomplishments make up the very origins of the mathematical systems we still use today. It began with our first perceptions of the plurality of objects and the geometry of shapes. The perception of three fish was different from two fish, and the same went for stones. Physical objects were perceived as long or symmetrical, rectangular or square. Eventually these perceptions were put into language: numerical signs and geometric drawings. In effect, counting began, as evidenced by ancient collections of pebbles, notches on pieces of wood, or knots in string. These pebbles, notches, and knots represent the first formal signs of mathematics.

Numerical Amounts and Geometric Structures

The archaeological record suggests that mathematical language became increasingly formal and abstract. In the Blombos cave of South Africa, two hundred miles from Cape Town, researchers have found two stones (dated to about eighty thousand years ago) that are engraved with geometric figures. At other sites in Africa, researchers have found bones inscribed with arithmetical calculations. A bone from the Lebombo Mountain caves in Swaziland (dating from thirty-five thousand years ago) bears twenty-nine notches that probably represent the number of days in a lunar cycle. Another bone showing calculations was found at the Ishango site in the Upper Nile Valley of the Democratic Republic of the Congo and is dated to twenty thousand years ago. The ancient Incas of South America also had a form of calculation, called *quipu*, that may have preceded their written language system. *Quipu* was a system of strings and knots that presumably was used by the accountants of the Incan Empire.

Relationships between Numerical Amounts

Today our mathematics is rooted in the idea of a set we call the natural numbers, designated by the notation $N = \{1, 2, 3, \ldots\}$. Sometime in the distant past, the relationships between these numbers were analyzed and put to good use. One of the first uses was the selection of a natural number to be the base, or basic unit, for calculations. In the modern field of discrete mathematics—which is popular in computer science and focuses on discrete structures that can be operated through finite processes—we can find proofs of the theorem that states that any natural number a can be represented by taking any other natural number b as its base. This is achieved by proving that the formula

$$a = r_n b^n + r_{n-1} b^{n-1} = \ldots + r_1 b + r_0 \text{ where } 0 \leq r_i < b$$

is valid for any two natural numbers a, b.

For example, we can write the number 109 in base 10:

$$109 = 1 \times 10^2 + 0 \times 10 + 9 = (109)_{10}$$

or in base 5:

$$109 = 4 \times 5^2 + 1 \times 5 + 4 = (414)_5$$

or in base 2:

$$109 = 1 \times 2^6 + 1 \times 2^5 + 0 \times 2^4 + 1 \times 2^3 + 1 \times 2^2 + 0 \times 2^1 + 1$$
$$= (1101101)_2$$

Likewise we could represent the number 109 in any base.

Before Greek civilization, we have no clear evidence of proofs of mathematical theorems. However, earlier accounts suggest that the idea of the base number was an aspect of intuitive knowledge that was put into practice. About 3400 BCE the Sumerians in Mesopotamia (today's southern Iraq) had developed a sexagesimal numeral system that they transmitted to the Babylonians. The sexagesimal system uses the number 60 as the base. Sixty has twelve factors $(1, 2, 3, 4, 5, 6, 10, 12, 15, 20, 30, 60)$, three of which $(2, 3, 5)$ are prime numbers. This system is very suitable for representing fractions. The twelve factors of 60 make it possible to divide the hour into half hours, periods of twenty minutes, quarters of an hour, periods of ten minutes, and so on.

In our modern world, the base 10 is most well known, a decimal system that arose from the number of fingers on our hands. However, from ancient times, the numbers 2, 3, 4, 5 on up were used as a base. In Egyptian hieroglyphs from 3000 BCE, such decimals are used to create very large numbers.

Calculations Based on Numerical Relations

The ability to calculate was the next step in primitive mathematics. A single calculation involves a mechanical procedure; we say it is mechanical because it can be executed by a machine, such as a computer. Children at school learn calculations in order to find the sum or the product of two numbers. The power of calculation can also be applied to great realms of complexity in the world, going from simple numbers to shapes in space. We can calculate the volume of a sphere simply by knowing the radius, the distance from the center to the edge. Even in primitive mathematics, some calculations were relatively complex.

The Moscow Mathematical Papyrus, so called for its home at the Moscow Fine Arts Museum, contains such calculations from Egypt in 1800 BCE. For example, problem number 10 on the papyrus calculates the area of a curved surface. Problem number 14 calculates the volume of a frustum of a pyramid on a square base with a height of 6, whose lower base is a square with a side measuring 4 and an upper base whose side measures 2. A diagram of this pyramid is shown in Figure 3.2.

The calculation of the volume given in the papyrus is $\frac{1}{3}6(16 + 8 + 4)$ and this responds to the general formula of the cal-

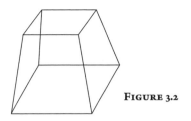

FIGURE 3.2

culation of a frustum of a pyramid $\frac{1}{3}h(a^2 + ab + b^2)$ where h is the height of the pyramid, a is the length of the side of the square of the lower base, and b is the length of the side of the square of the upper base.

Another example of early calculations is found in the Rhind papyrus, an Egyptian document from 1650 BCE obtained by the Scotsman Henry Rhind (also called the Ahmes papyrus). One segment of the papyrus is housed at the British Museum in London, and this specimen reveals that the ancient authors knew how to decompose a number into its prime factors and how to present arithmetical and geometric series. The papyrus also contains an approximate calculation of $\pi \approx 3.16$, that is, the formula for pi, which is the ratio of the circumference of a circle to its diameter ($\pi \approx 3.14159265+$).

Intuitive-Empirical Theorems

Before Greek civilization, knowledge of mathematics was empirical. These earlier cultures, based on an intuitive sense of mathematical relations made empirical calculations that were correct, despite the lack of a formal mathematical language. Well before the proof of important theorems attributed to the Greek thinker Pythagoras, sages in China, India, Babylon, and Egypt understood aspects of what came to be known later as Pythagorean triples. These triples, for example, are threesomes of whole numbers (3, 4, 5), (5, 12, 13), (7, 24, 25), and so on, that define the sides of a right-angled triangle, such that the sum of the square of two of them equals the square of the third, written as $a^2 + b^2 = c^2$. For example:

$(3, 4, 5)$ verifies $3^2 + 4^2 = 5^2$;

$(5, 12, 13)$ verifies $5^2 + 12^2 = 13^2$;

$(7, 24, 25)$ verifies $7^2 + 24^2 = 25^2$ ...

Before Pythagoras developed the deductive means of proving this theorem, earlier cultures had arrived at knowing its reality by induction—that is, by experience and experiment. We can almost imagine the ancient thinkers putting objects, such as squares and triangles, on the ground and looking at how their sizes, angles, and lengths compared. This was the experimental (inductive) study of mathematics.

The Infinite in Mathematics and Religion

As discussed in the previous chapter, the infinite is perhaps one of the most basic human intuitions, and it naturally played a role in the rise of primitive mathematics. The pre-Greek epoch offers evidence of testimonies of the use of the infinite that carried both religious and mathematical connotations. In the eighth century BCE, the Hindu text of the Yajur Veda used the concept *purna*, or "fullness," to state that if we subtract *purna* from *purna*, what remains is *purna*. Many centuries later, the mathematician Georg Cantor (d. 1918) developed a theorem to prove this idea of infinite sets.

In other words, the concept of *purna* has been interpreted both religiously and mathematically. As we noted earlier, we can avoid confusion between the religious and mathematical infinite by a clear distinction between formal signs and metaphysical symbols. Faith is expressed by saying that the transcendent God is infinite because God has no limits. Symbolically, this statement means that we see the world as limited and God as unlimited.

In mathematics we state that a set is infinite if we cannot finish counting elements—something the ancients understood quite well. When counting the set of all natural numbers—N = {1, 2, 3, ...}—we never finish. Here, the infinite stands for a formal concept, the set {1, 2, 3, ...}. Despite the important distinction between the metaphysical and mathematical infinite, some overlap exists as well, because both deal with a great unknown. Naturally, the ancient founders of mathematics were fascinated by both kinds of mysteries, and in Greek and medieval mathematics we again see this interplay of the formal and the religious—as the next chapter shows.

CHAPTER 4
Euclid and Beyond

THE BLOSSOMING of Greek thought, beginning in the sixth century BCE, has typically been recounted in the lives of Socrates, Plato, and Aristotle. But mathematics was also a key component of this remarkable period, and what the Greeks put down in writing might have been lost, had Roman and Arabic chroniclers not rescued some of these materials. After the rescue, the materials surfaced again in the twelfth century CE in Europe, stimulating a renaissance in mathematical exploration after roughly fifteen hundred years.

The chief characteristics of the early Greeks' contribution to logic and mathematics may be described as the development of theoretical proofs. Eventually these proofs were applied to deductive science, in which a science begins with a first principle and then tests that premise against the varieties of phenomenon in the world. In the medieval period, deductive science was expanded in its application to algebra. In both the Greek and medieval periods, as we will see, the formalism of math invariably overlapped with the religious culture of the times.

The great transition of mathematics from trial and error to a deductive science came with the work of three primary thinkers: Thales of Miletus (d. ca. 546 BCE), Pythagoras of Samos (d. ca. 500 BCE), and Euclid of Alexandria (d. ca. 256 BCE). The kinds of new mathematical proofs that we attribute to Thales and Pythagoras, for example, likely have roots in Babylonian and Egyptian culture. However, the Greeks formally moved from intuition and

simple calculation to using first principles—called axioms or postulates—as the starting assumptions needed to prove the validity
of mathematical theorems.

By making mathematics deductive, the Greeks established a sta-
ble point from which to explore complex mathematical realities
that otherwise would boggle the mind. The most significant example of this structural stability is Euclid's treatise on mathematics,
the *Elements*. Here we find a compendium of proofs of arithmetical and geometric theorems based on axioms and elementary principles.

Until that point, all math was based on calculations that followed mechanical rules, which, when applied to some data, produced other data. The calculation of the square root applied to
the number 25 produces the number 5. Not much has changed
in the nature of calculations such as this. Because they are strictly
formal processes, these rules amount to algorithms—a mechanical sequence—that today we put into language that a computer
understands.

What, then, is the difference between a proof and a mere calculation?

Whenever we try to use reason to convince someone of the truth
of a proposition, we are reasoning informally. To do so, we base our
reasoning on propositions that we suppose are valid for the person
addressed and that make it possible for us to deduce the proposition we wish to prove. When applying for a job, for example, an
applicant puts forward arguments in the form of a résumé, hoping
that the company accepts the résumé as the premise—and proof—
of the job seeker's suitability for the position.

Likewise, in mathematics we try to prove a proposition based
on given axioms, which operate like the résumé: it is the bedrock
on which our arguments build. So Thales, Pythagoras, and Euclid
attempted to prove the validity of certain mathematical propositions based on the validity of other clearer and more evident propositions. In the *Elements* (book 7, 1–3), Euclid proves that we

have a type of calculation that finds the largest common divisor of two numbers. Given any two natural numbers, in other words, the Euclid algorithm calculates the largest number that can divide both.

Euclid's *Elements* served for centuries as a basic mathematics text. Until the nineteenth century, in fact, people thought that the geometry deduced from Euclid's axioms was the only geometry possible. Today we know non-Euclidean geometries whose propositions are deduced from different sets of axioms. We also know that Euclid's geometry was presented in a rather informal language; twentieth-century mathematicians assumed the task of making that language formal in terms of mathematical signs and syntax.

We can see how Euclid conveyed his proofs in the example of the Pythagorean theorem about the sides of a triangle (*Elements*, book 1, 47; see Figure 4.1). Here, Euclid proved that, given a right-angled triangle ABC, the area of the square BCDE, constructed by taking the hypotenuse of the triangle as a side, equals the sum of the areas of the squares ABGF and ACHK, constructed taking as sides the other two sides of the triangle. Euclid built his proof on previous axioms and propositions. So to prove his proposition 47 (regarding squares BCDE, ABGF, and ACHK), he builds upon proposition 46 (regarding a line and square) and upon axiom 1, which defines a line as passing through two points.

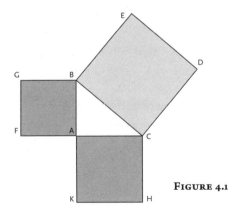

FIGURE 4.1

When we say that Euclid was using an informal language, despite his careful building up of proofs from deductive logic, we can illustrate this limitation by comparing Euclid to our computer age. Euclid's language lacked formal rules of deduction as well as the formal language (signs) in which he could write his proposition. As a result, Euclid's profound mathematics cannot be programmed into a computer today, unless we translate it into a formal language, since his proofs lack the necessary mechanical quality.

Let's look more closely at Euclid's informal language. In his proof of the Pythagorean theorem. he explains this logical rule: if we know that proposition B is true and that proposition A can be deduced from B, we can always state that A is true. This logical rule is usually called *modus ponens,* or more precisely, *modus ponendo ponens* (mode that affirms by affirming). It can be written down formally as:

$$\frac{B, B \to A}{A}$$

This use of logical evidence is normal in mathematical proofs, but it is also the way we argue a case in everyday life. In contrast, we can look at the way a computer makes arithmetical calculations based on formal rules, the kind that Euclid did not use in his groundbreaking work. As seen in Figure 4.2 on the next page, with a formal language of rules, a computer can arrive at a mathematical proof about triangles by a mechanical calculation.

Calculation of addition:

$\{3, 5\} \to \{\text{Sum calculation rules}\} \to \{8\}$

Calculation of the product:

$\{3, 5\} \to \{\text{Product calculation rules}\} \to \{15)$

Calculation of the theorem of Pythagoras:

$\{\text{Geometry axioms}\} \to \{\text{Logic rules}\} \to \{\text{Theorem of Pythagoras}\}$

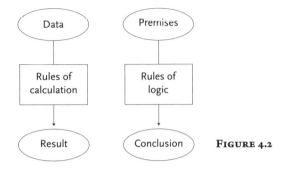

FIGURE 4.2

Euclid hinted at such rules but did not formally recognize them as we know them today. As Euclid seemed to understand, the human mind responds to this kind of logical argumentation. If Euclid had been much more specific in writing down the rules that governed this logic, they would become much more difficult to consider. Ordinary people would have lost the spontaneous ability to understand Euclid's logic. But while the human mind can be slowed down by having to contemplate a deeper set of mechanical rules, a computer does not have this problem at all. Indeed, a complex set of mechanical rules helps a computer do more. Computers need precise rules on what to do (logically) in each case. We know complete logical systems with which we can formalize all the mathematical proofs that, once formalized, can assume the form of a procedure, which a computer can then "understand."

Of course, we are not detracting from Euclid by making this comparison. For two thousand years and more, the logic of the *Elements* matched perfectly the capabilities of the human mind, allowing for the rise of more advanced geometry and eventually modern science.

THE RISE OF FORMAL LOGIC IN GREECE

There are two Euclids in the history of mathematics, and we should not confuse the two as we try to look more closely at the origins

EUCLID AND BEYOND : 49

of formal logic in Greece. The setting for this development was the expanding school of Stoic philosophy, which began in the fifth century BCE. Euclid of Megara founded one of these early Stoic schools and also developed the logical system that is the basis of propositional calculus. (We briefly consider propositional calculus, the basic logical structure in human language, in chapter 7.) Later, the school of Stoic philosophers founded by Zeno of Citium (d. 264 BCE) continued to develop the formal propositional calculus. Aristotle (d. 322 BCE) described some formal inference rules that go beyond propositional calculus, but did not carry out a methodical and complete development of the logic language as we understand it today.

The formal mathematical language of today—using signs— began with the algebraic studies of George Boole (d. 1864), followed by the development of formal languages by Gottlob Frege (d. 1925), Kurt Gödel (d. 1978), and other mathematicians. With these accomplishments, mathematicians set down formal rules of inference and logic. But without the groundwork of the Greeks, this task would have been more difficult. Even these great modern mathematicians began their careers pondering Euclid and the others.

On the road to our modern-day formalization of mathematical language, a major discovery was algebraic equations. Algebra is a system that uses symbols, such as x and y, to stand for numbers or other mathematical objects that can be inserted to carry out calculations. For most students in the modern world, algebra is the highest form of math that is studied in a general education: it is a rudimentary introduction to the signs of formal logic and math.

This algebraic approach has precedents among the Egyptians and Babylonians, who worked with equations that included unknowns to be solved. Since they had no formal language for this, however, something like modern algebra did not emerge. Around 250 BCE, the Greek thinker Diophantus of Alexandria was among the first to introduce formal signs to write polynomial equations (equations with two or more terms). These are now written as

$a_n x^n + a_{n-1} x^{n-1} + \ldots + a_2 x^2 + a_1 x + a_0 = 0$. Diophantus used this method to resolve linear and quadratic equations in a language that was still partly informal. His equations only sought solutions for positive and rational numbers. He did not consider negative or irrational solutions; thus these numbers were outside of mathematics. (Appendix 5 describes the number systems in more detail.)

MATHEMATICS AND METAPHYSICS IN GREECE

Our best examples of how mathematics and metaphysics mingled in Greece are the schools of the Pythagoreans and the Platonists. Both of these schools gave mathematics a symbolic metaphysical meaning that went beyond formal mathematical language. We know little about Pythagoras, but the outlines of his school and disciples are quite clear. Based on the objective discoveries that Pythagoras made regarding mathematics, his enthusiasm inspired his disciples to found a community that sought religious significance in numbers. However, this metaphysical enthusiasm came up against a number of logical paradoxes.

The Pythagoreans found in numbers the last meaning of reality. By taking this absolute and mystical stance, they ended up confusing mathematics and metaphysics. Aristotle mentioned this mystical school in his *Metaphysics* (book 1, chapter 5): "Pythagoreans supposed the elements of numbers to be the elements of all things, and the whole heaven to be a musical scale and a number." Aristotle traced this metaphysical enthusiasm to the discovery of numerical concordances between musical intervals and the movements of the stars. He was deeply skeptical, however, as shown by his further comments on how the Pythagoreans operated:

And all the properties of numbers and scales which they could show to agree with the attributes and parts and the whole arrangement of the heavens, they collected and fitted into their scheme. . . . If there was a gap any-

where, they readily made additions so as to make their whole theory coherent. E.g. as the number 10 is thought to be perfect and to comprise the whole nature of numbers, they say that the bodies which move through the
· heavens are ten, but as the visible bodies are only nine, to meet this they invent a tenth—the "counter-earth."

The Pythagoreans endowed the first ten numbers with mystical properties and believed that 10 itself was most sacred. The Pythagorean Tetraktys (see Figure 4.3) were based on the relationship 4 + 3 + 2 + 1 = 10. (The "tetra" stands for the four rows that organize the ten points into a triangle, which became a mystical symbol for the Pythagoreans.) They also gave esoteric significance to the pentagram based on the curious numerical relations between its segments.

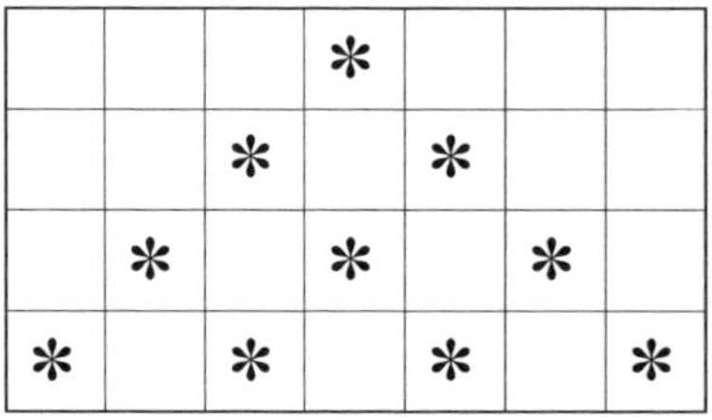

FIGURE 4.3

THE PYTHAGOREAN TETRAKTYS

The Babylonians had already given symbolic value to the pentagram. The Pythagoreans cultivated this tradition and represented the pentagram with its points directed upward (see Figure 4.4). The pentagram has had a universal appeal. Later, Christians used it to represent the wounds of Jesus. Moreover, the neopagan Wicca religion has adopted the pentagram as a mystical symbol.

Pythagoras's most significant mathematical achievement was his deduction that for any right-angled triangle whose two legs have a length equal to 1, the length of the hypotenuse will be $\sqrt{2}$. However, since we cannot represent the value of $\sqrt{2}$ as a quotient

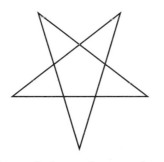

FIGURE 4.4

of two whole numbers, or through a finite succession of numbers, we must conclude that the diagonal of a square and the side cannot be measured with a common unit. The lengths that cannot be measured with a common unit are called incommensurable. The numbers that can be represented as quotients of two whole numbers, and therefore measured with a common unit, are called rational numbers. One member of the Pythagorean school, Hippasus of Metapontum, discovered the fact that √2 is incommensurable and not a rational number. (Appendix 5 includes a proof that √2 is not a rational number.) Whatever the case, the Pythagoreans adopted the religious dogma which said that whole numbers and their fractions are the essence of the universe.

Platonism

Platonism is the broad and varied philosophical system that has been based on interpreting the many works of Plato (d. 347 BCE), the Athenian who was a student of Socrates and the teacher of Aristotle. Unlike Aristotle, Plato believed that the essences of the universe were found in numbers. The Platonist view of mathematical objects has recurred across history. It is a form of mathematical realism, according to which mathematical objects are eternal and unchanging realities; they exist as realities (realism) outside the human mind's own ideas (idealism). Down to the present, Platonist mathematicians argue that we perceive these eternal mathematical realities as intuitions.

In his *Republic,* Plato presented his model of eternal Ideas or Form by using the allegory of the cave. In this story, prisoners chained in a cave see shadows of people passing by the fire and believe that the shadows are forms in the real world. However, Plato explains, the philosopher is like a prisoner who is freed from the cave and realizes that the shadows are mere copies of a higher reality. That higher reality is home to eternal essences, which includes absolute and ideal numbers. Plato's mathematical realism has had a lasting influence. The belief that mathematical objects are independent of our mental invention is the basis of classical mathematics. In contrast, the constructivist school argues that mathematical objects are true only as useful and logical constructions of the human mind.

THE MEDIEVAL PERIOD

With the loss of so much of the Greek legacy as recorded in manuscripts, creative activity in mathematics was delayed in Europe for many centuries. In the meantime, however, mathematical contributions were being made in China and India, and these began to meet each other in the context of Islamic culture.

For the early medieval period, Islamic culture was a kind of ecumenical stimulus for scientific development, to which mathematicians from the Muslim, Greek, and Indian worlds all contributed. In particular, this development was made possible in Baghdad around 800 CE under the aegis of Hārūn al-Rashīd, the fifth caliph of the Abbasid dynasty. He invited scholars from various regions to come to Baghdad and do scientific research. From India came three significant contributions to the new mathematical synthesis: the decimal system, negative numbers, and zero. At this time also, Euclid's *Elements,* written in Greek, was translated into Arabic. Moreover, the ancient formalism that Diophantus of Alexandria used to resolve several types of polynomic equations—today written as $a_n x^n + a_{n-1} x^{n-1} + \ldots + a_2 x^2 + a_1 x + a_0 = 0$—was revived and applied to the new medieval mathematics.

The Islamic world had its own mathematical innovators as well. In Baghdad, Abu Abdallah Muhammad ibn Mūsā al-Jwārizmī (or simply al-Jwārizmī) systematized the new materials and originated the words "algebra" and "algorithm." Our pronunciation of "algorithm" follows al-Jwārizmī's name. Similarly, our pronunciation of "algebra" reflects the title of his treatise (*al-Kitāb al-mukhtasar fī hisāb al-jabr wa'lmuqābala*), which was translated into Latin in the twelfth century as *Algebra et Almucabal* (for *al-jabr wa'l-muqābala*).

Just as the Greeks introduced into mathematics the process of proving theorems, Islam gave a strong boost to algorithmic calculations. An algorithm, as mentioned earlier, is a procedure that can be executed mechanically—a succession of instructions that can be run automatically with no need to make a decision. Cultivated in Islam, algorithmic methods began to circulate in Europe in the thirteenth century. The Italian mathematician Leonardo Pisano (d. 1250), also known as Fibonacci, packaged and disseminated a decimal number system with zero and nine digits, a system he learned while traveling in North Africa. In 1202 he published the *Book of Calculations* (*Liber Abaci*), which spread these methods across Europe.

Our final question for the Middle Ages is how the new mathematics overlapped with the metaphysics of Christian and Islamic thought. Simply put, the idea of the infinite was a central topic for both mathematical calculations and the philosophical arguments about the nature and attributes of the Creator. Second, just as mathematics became a science of proofs from axioms and calculations, medieval theology, both Christian and Islamic, was expansive on the topic of proofs for the existence of God.

CHAPTER 5
Dawn of Science

IN THE SEVENTEENTH CENTURY, mathematics emerged as a language for talking not just about ideas in the mind but also about the laws of nature. This began with physics—first looking at objects on earth and then heavenly bodies—and eventually extended to chemistry and biology. Today, the language of mathematics dominates even our social sciences—that is, our human conduct as studied by sociology and economics.

This chapter emphasizes how mathematics and empirical science began to augment each other, and also how mathematics and metaphysics interacted. In this period we have much better biographical material about the scientists who moved this process forward, so we can use their stories as a way to chart these radical new developments in mathematics.

GALILEO AND THE PHYSICS OF ARISTOTLE

None of these historical figures was more colorful than Galileo Galilei (d. 1642), a native son of Pisa, Italy. Galileo used scientific instruments to observe nature and described his observations in mathematical language. This quantitative use of mathematics distinguished him from earlier figures, who did not use mathematics to describe empirical knowledge.

Galileo conducted his scientific observation at a time when Aristotle's theory, presented in his *Physics*, dominated science and theology. Aristotle's popularity put Galileo on a collision course with the conventions of the time, and this disagreement with Aristotle was at the heart of Galileo's final clash with Roman Catholic authorities on whether the earth moved or not. Aristotle argued that the earth was stationary and anchored the universe.

But for the development of modern science, the chief theory that held back scientific advance was Aristotle's definition of the four causes of physical motion: the material, formal, efficient, and final causes. In summary, Aristotle argued that every type of thing has a particular nature, and this nature determines its movements, directions, and purposes. However, scientists including Galileo questioned these Aristotelian assumptions. Galileo did so because, through the use of experiment and observation—most famously with the telescope—he saw that the universe did not seem designed the way that Aristotle claimed. Nor did the causes seem to be helpful explanations of how material objects really moved in space.

Recall that Aristotle was not enthusiastic about mathematics, and it was a point of criticism he held against his teacher Plato. Galileo would side with Plato in this sense, realizing that mathematics was the language of nature. As Galileo eloquently stated in his writings,

Philosophy [i.e., physics] is written in this grand book— I mean the universe—which stands continually open to our gaze, but it cannot be understood unless one first learns to comprehend the language and interpret the characters in which it is written. It is written in the language of mathematics, and its characters are triangles, circles, and other geometrical figures, without which it is humanly impossible to understand a single word of it; without these, one is wandering around in a dark labyrinth.

When Galileo made his quantitative measurements of falling, swinging, or rolling objects, and expressed them in mathematics, he excluded all of Aristotle's causes. By his use of the language of mathematics, Galileo suggested our explanations of physical phenomenon could now unify all causes. Being able to describe cause and effect precisely is at the very heart of science. This new focus of science, however, rejected all talk about ultimate causes, looking only for direct causal relationships.

One example of this direct causal relationship is the acceleration of a body. On observing how a body of a certain mass accelerates, we can state the physical law at work by the equation $f = ma$: force (F) is the product of the mass (m) and acceleration (a). According to this law, an increase in the acceleration of a given body causes an increase in force; in turn, the increase in force causes an increase of the acceleration. The cause and effect goes both ways, and this is how a modern mathematical formula such as $f = ma$ is understood in modern physics, quite contrary to Aristotle's view.

Another way to say this is that Aristotle's four types of causes did not describe how things behave, but rather how things are. As a result, Aristotle made many serious errors in describing the conduct of things. He believed, for example, that if one body weighed more than another, it would fall more rapidly because the very weight of an object is determined by its inclination to occupy a final place, such as falling to a state of rest. Galileo tested this, however, and found that, in theory, all objects fell at the same rate (if we take away the resistance of air). Arguing that the distance a body falls is proportional to the square of the time elapsed during the fall, Galileo used a mathematical notation such as this: t^2. Hence, the mathematical and physical causes of the speed a body travels a certain distance is the square of the time elapsed.

On many occasions Galileo carried out his observations with very simple means. To measure time elapsed, Galileo weighed the water emerging from an orifice. Two periods of time were equal if the same amount of water emerged during these periods. To control the speed at which objects fell, Galileo rolled balls down an inclined plane. This allowed him to measure a slower version of a ball falling, which would be much faster if it fell from the sky. (See Figure 5.1.)

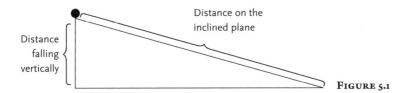

Distance on the inclined plane

Distance falling vertically

FIGURE 5.1

Let's look at how Galileo used mathematics to measure this law of motion, distance, and time. First he placed horizontal wires at different heights along the inclined wooden plane. These wires clicked when a ball passed over them. Galileo moved the wires up and down the plane until the time elapsing between two clicks was the same. He observed that if the time elapsed between two clicks was the same, the distance traveled between these clicks increased in accordance with the time elapsed. The mathematics goes as follows:

Distance traveled from commencement to the first click:
$d = k \times 1^2$

Distance traveled from commencement to the second click:
$d = k \times 2^2$

Distance traveled from commencement to the third click:
$d = k \times 3^2$

In other words, Galileo actually observed that the distances traveled were proportional to the square of the time elapsed. This represents the emergence of the modern scientific method of empirical observation, followed by representation in mathematical language. Since the time of Galileo, the methods and instruments of observation and measurement have multiplied and diversified. But the method behind all of them was present when Galileo studied balls rolling on inclined planes.

Galileo's use of the telescope is also prototypical of modern science. The optical lens had been invented outside Italy, and the telescope was more of a curiosity than a scientific tool. However, Galileo realized its use in making astronomical observations, and he capitalized on this in his quest for success and fame, and to popularize science. In the process, he advanced the scientific method.

As a first step, he observed mountains and craters of the moon. From this he concluded that the heavenly bodies cannot be perfect geometrical objects as Aristotle and Plato had long suggested. In Galileo's day, the standard astronomical model came from the second-century Hellenistic scientist Claudius Ptolemaeus of Alexandria. The Ptolemaic model had already been challenged by the work of Nicolaus Copernicus (d. 1543) and Johannes Kepler (d. 1630), but because their theories were purely mathematical, they did not spread and create controversy. A century after Copernicus, however, Galileo, based on his physical observations, argued that the earth moved around the sun—and publicized his ideas openly.

These discoveries began on January 7, 1610, when Galileo observed three moons rotating around Jupiter. After several nights of further observation, he found that four moons, in fact, moved around Jupiter. In March he published a short treatise in Latin with the title *Sidereus Nuncius* (*The Heavenly Messenger*) in which he described his telescopic observations of the moon, the stars, and the moons of Jupiter. In particular, the observations of Jupiter contradicted the traditional view (of Aristotle and Ptolemy) that all heavenly bodies orbited around earth, which was stationary in the universe.

The Changing Relationship of Science and Religion

Galileo's discoveries also marked a final break between science and the traditional authority of religion and philosophy. In this break, science gained autonomy. Its only authority was its own observation and quantitative measurement. The truths of science were declared in mathematical formulations that tried to describe laws of nature. The point had been made before Galileo, but his words about a "grand book" of nature, inscribed with the language of mathematics, remained a powerful image of why nature is intelligible to us.

This joining of physical observation and mathematical language nevertheless lives in tension, which means that, internally, science remains perpetually in tension as well. As time has passed, the tension eased somewhat as mathematical language was formalized and the very instruments we use to measure the physical world came to be designed and calibrated according to mathematical formulas. But observation and mathematical formulas do not always match up so easily.

A tension has continued between science and religion as well, although clearly science had declared its autonomy, and for very good reasons. Science can remain autonomous because it uses mathematical language based on formal signs. These signs only claim to represent observed data. In general, science makes no claims over the symbolic kind of language used in metaphysics or religion. The symbols of religion, in turn, cannot be reduced to quantitative signs. This is a healthy distinction between science and religion.

For better or worse, Galileo had to come in conflict with the church to make this distinction. As his conflict deepened, he tried to articulate this kind of separation—saying, for example, that the Bible does not state how the heavens go, but instead how to go to heaven. In a bit of desperation, perhaps, he even cited passages in the Bible to suggest the scientific fact that the earth could move. Although he was a strong-minded innovator, Galileo was also a loyal Catholic but one who was willing to clash with the theologians. At the time, the theologians followed Aristotle, and the top Vatican official on the Galileo case, Cardinal Robert Bellarmine (d. 1621), was worried that the outspoken scientist would undermine the faith of the public by radically changing the biblical (and Aristotelian) view of the cosmos.

In the end, Galileo faced two church trials. The first came in 1616 and was based on his writings that the earth moved around the sun. The judge in this case was Bellarmine. The cardinal told Galileo that he could only speak of his heliocentric theory as a hypothesis and that he must abstain from attacking the church's stance on cosmology. However, Galileo did not abstain, which led to his second trial in 1633. This trial resulted mainly from a power struggle between the Aristotelian view and the modern view of science, and between the groups that had vested interests in either one. When Galileo was sentenced to house arrest in 1633, the Aristotelian camp was victorious. In 1642, at his well-appointed home in Arcetri, near Florence, he died peaceful and comfortably, but in scientific exile. On October 31, 1992, almost 360 years later, Pope John Paul II acknowledged that the trial of Galileo had been unjust and had damaged efforts to reconcile science and religion.

MATHEMATICS AS A LANGUAGE OF EMPIRICAL SCIENCE

The work of Galileo was only an early step in translating physical observations into mathematical formulas. Other scientists followed this path, and Englishman Isaac Newton (d. 1727) finally crystallized the most basic physical laws into a simple set of mathematical representations. He did this in his three-volume treatise *Philosophiæ Naturalis Principia Mathematica*, probably the most important document in the unification of empirical and mathematical science. The *Principia* contains Newton's universal law of gravity, the law of classical mechanics—the only mechanics known until the theory of relativity—and the laws governing the movement of the planets, which were discovered by Johannes Kepler and incorporated by Newton.

Newton's law of universal gravity is an example of an empirical law formulated mathematically. The law establishes that the force (F) whereby two bodies (m_1 and m_2) are attracted is proportional to the product of their masses; the force is also inversely proportional to the square of the distance between two bodies. The mathematical formulation is as follows:

$$F = G\frac{m_1 m_2}{r^2}$$

where G is a gravitational constant and r the distance between two bodies. The law of universal gravity is a mathematical theory that explains the movements of all the bodies under the influence of gravity and, in particular, the movements of the planets in the solar system.

The law of universal gravity is more than a mere mathematical proposition, for though it can be written with mathematical signs, it represents observations of reality, not simply intuitions of what is logical. The law of universal gravity is a scientific hypothesis that can be tested, and as long as it passes the empirical test, it will be called a solid scientific theory of how the world really works.

As is well known, Newton's theory described the broad pattern of the solar system but could not explain some anomalous details, such as the odd way Mercury's orbit came very close to the sun (called the perihelion, or nearest point). This was one of the problems that Einstein's theory of relativity in 1915 set out to solve, and by building on Newtonian physics, Einstein's explanation made his radical new theory acceptable to science.

Newton offered one other great advance to the mathematics of modern science, and he did so in a colorful feud with an equally brilliant mathematician, German philosopher Gottfried Wilhelm Leibniz. The advance they both contributed to was the fundamental theorem of calculus, which produced the mathematical tool of infinitesimal calculus.

With infinitesimal calculus, we can measure continuously varying magnitudes, allowing us to calculate the infinitesimal variations of a function (that is, a differentiation) that gives rise to surfaces or volumes. In short, calculus has become indispensable for describing the mathematical conduct of the laws of nature. Although Leibniz and Newton discovered the fundamental theorem of calculus independently, Europeans hotly debated who was first (or whether the idea was stolen!). Either way, the English naturally sided with Newton and the continentals with Leibniz, and each camp adopted the notational system of its favored teacher. Today, modern science relies on infinitesimal calculus as indispensable for describing the laws of nature and even for predicting the arcing flight of spaceships to the moon or other planets.

Modern Mathematics and Metaphysics

In the centuries before the age of Galileo and Newton, the Pythagoreans, the Platonists, and the medieval theologians projected the rational mathematics of the Greeks into metaphysical systems. In the modern era of mathematics, three new kinds of metaphysical arguments rose to prominence: the principle of sufficient reason, the belief in causal determinism, and the rejection of metaphysics. Paradoxically, the rejection of metaphysics is metaphysics itself, when this rejection is justified with arguments about the ultimate causes of things.

In addition to being a mathematician, Leibniz was also a rationalist philosopher in the lineage of René Descartes and others on the European continent. (We recall that Leibniz, like Descartes, accepted a version of the ontological proof for God's existence.) Leibniz is also famous for explicating the idea that nothing occurs in nature that does not have a sufficient reason to explain the occurrence. The principle of sufficient reason is equivalent to affirming that if God knows all the causes, and the reasons such causes have for occurring, then God knows the reasons why anything happens. At a time when mathematics was trying to eliminate ultimate causes, Leibniz's system sounded like fatalistic determinism, which would deny human free will and all chance and creativity in nature.

In order to salvage human free will, Leibniz distinguished between people's limited knowledge, which enables them to take decisions freely because they do not know all the reasons why anything happens, and the unlimited knowledge of God, who knows all the causes or reasons.

For Leibniz, everything in the world followed a plan of preexisting harmony that was clear in the mind of God. Hence, God had created the best possible world. All aspects of this Leibnizian world, natural and supernatural, are linked together and can be sufficiently understood by rational methods modeled on mathematics. In talking about the world, Leibniz believed that all men of good will could solve problems together by saying, "Let us make the calculations." As advanced and sweeping as Leibniz's vision seemed, for his system of a harmoniously deterministic universe he still relied on the basic logic provided by the Greeks.

The French scientist Pierre-Simon Laplace (d. 1827) took the idea of sufficient reason even further, beginning his major work on determinism by citing Leibniz. Laplace, however, seemed even more deterministic than his German predecessor. Laplace said that if everything has a mathematical explanation, everything would be deducible from certain basic mathematical propositions. He was also the successor to Newton in analyzing the solar system, believing that, based on Newton's laws, all phenomena could be predicted based on the location and momentum of the atoms making up matter. Laplace explained his determinist view by saying that if there were a demon (or an ultimate Mind) that knew the position and momentum of all the atoms in the universe at any given moment, this demon could predict the future in utter detail. Whereas Leibniz attributed the determinism of the world to God's mind, Laplace apparently concluded that modern science no longer needed the God hypothesis to explain the workings of the world.

The third consequence of the belief that everything has a mathematical explanation is the complete rejection of metaphysical knowledge. We see this suggested in Laplace, but the English philosopher David Hume (d. 1776) made the assertion even more explicit in his 1748 essay *An Enquiry Concerning Human Understanding*. The essay is a spirited attack on all metaphysical thinking. It is also a manifesto for modern empirical science as the only true form of knowledge. In taking this stance, Hume concludes (chapter 12, part 3) with these words:

> When we run over libraries, persuaded of these principles, what havoc must we make? If we take in our hand any volume of divinity or school metaphysics, for instance, let us ask. Does it contain any abstract reasoning concerning quantity or number? No. Does it contain any experimental reasoning concerning matter of fact and existence? No. Commit it then to the flames, for it can contain nothing but sophistry and illusion.

In Hume's eagerness to burn all metaphysical writings as useless, many subsequent philosophers tell us, Hume himself had taken a metaphysical stance of his own. He has pitted his metaphysical— that is, ultimate—view of the universe against the metaphysical view that religions typically hold. As ever, metaphysics refers to ultimate principles, not to a methodology such as scientific empiricism or mathematical logic.

Different cultures have arrived at different metaphysical visions, and as we have seen, mathematics or religion can aid in this stance. In the modern age, science itself has also offered a metaphysical vision. This vision says that our physical senses can apprehend all that exists. Of course, this remains an open question. Is human knowing truly limited to what can be quantified? With the writings of Hume we see the advent of positivism (or philosophical materialism), one of the most influential metaphysical systems of our modern age. But as we will see, further advances in mathematics suggest that science must remain open and plural on such ultimate questions. It seems premature to declare, along with Hume, that there exists one positivist universe that everyone can agree upon; even today, mathematicians, scientists, and religious thinkers hold a variety of beliefs about ultimate things.

CHAPTER 6
Mathematics Formalized

UP TO THE twentieth century, mathematics had been gathering every kind of language that was helpful to its work. The first mathematicians put notches on sticks and knots in strings, while centuries later they began to talk about their ideas in shapes they inscribed on papyrus and in their natural languages of Greek, Arabic, Latin, Chinese, and Hindi. Eventually, they began to agree on formal signs that could be used in arithmetic, geometry, and algebra—signs such as $+$, $-$, $\times$, $\div$, which stood for sums, differences, products, and divisions. Letters such as "a" and "b," moreover, were used to represent unknowns in mechanical calculations such as $a + b = b + a$.

The final frontier of mathematics, however, was to create a comprehensive and formal system that had enough signs and rules to speak of every possible relationship, and to make this system coherent and without contradiction. This was the achievement, for the most part, of the twentieth century, although we must say it is not a utopian achievement (given the absence of universal agreement on the best formalized system).

The formalization of the language of mathematics has had two important consequences that we review in this chapter. First, such a formalization allows us to study mathematical language itself by using mathematical methods. Just as mathematics proves theorems about numbers, straight lines, planes, geometrical figures, and other mathematical objects, it now can be used to prove the validity of language and meaning of a mathematical system. This is mathematical self-knowledge; it has allowed us to evaluate decisively the certainty of mathematical propositions.

The second consequence has been truly revolutionary for our practical lives in the modern world. The formalization of mathematical language has produced an automation of mathematical proofs. This automation has connected the human mind with the computer. Once we formalized mathematical language, it could be put in computer language, and computers can calculate and analyze data with a speed and power that are not possible in ordinary human thinking.

These two consequences take us beyond what mathematics had done in the way of providing representational language for the natural sciences. It is not uncommon today, for example, to see natural scientists—astronomers, biologists, and even social scientists—leaving the natural world and going into a laboratory to spend months and years running mathematically driven computer models to try to understand nature and human beings. At the same time, the great debate among professional mathematicians is the validity and coherence of entire mathematical systems. In short, formal mathematics has declared a kind of autonomy from even the empirical sciences—a development that came about gradually with great milestones in mathematics across the twentieth century.

Truth and Certainty

Modern mathematics burst onto the contemporary scene with troubling new questions about the very truth and certainty of older mathematical approaches. In the past, certainty was based first upon intuitions and second upon axioms and proofs. But the consistency of these older certainties began to fall into doubt. Everyone knew that mathematics produced plenty of logical paradoxes, but this fortress was not attacked until the rise of modern mathematics. So the quest began to create a formal system that avoided all paradoxes. Such a system aimed for complete consistency.

This quest for certainty in mathematics had a strong metaphysical ring, of course. One goal was to find a kind of absolute, unchanging truth. As we see later in this chapter, this is the most significant overlap between science and metaphysics in the twentieth century. To find an absolute system, mathematics has tried to reduce all truth to formal signs. This approach would have a great impact on the more general discussion of how science and religion relate to each other. The modern idea of truth has come to hinge on the notion of mental consistency, and to the extent that we can agree that consistency is a kind of certainty, science and religion can speak to each other of their own internal consistencies and certainties.

Consistency, Completeness, and Decidability

When modern mathematics began to study the nature of its own systems, it was forced to ponder three new questions: Was there consistency in a system? Was a system absolutely complete, accounting for all possibilities? And finally, could a system guarantee a decision on any formula, known as the problem of decidability. The second and third of these questions—completeness and decidability—have turned out to be the most problematic for modern mathematical systems. The great minds of the twentieth century had to conclude that some systems cannot be complete and some mathematical questions cannot be decided. The implication is that truth and certainty are limited, and since mathematics underlies much of science, that uncertainty also extended to the scientific world. As this chapter shows, however, the enduring existence of consistency finally gives our modern age its certainty in math, science, and metaphysics.

In formalized mathematics, we speak of certainties as either univocal or plural. A univocal certainty can be expressed in the same way in all contexts and in all cultures. From the point of view of their formal expression, univocal certainties are the same for everyone regardless of gender, social class, or religion, since a formal method can illustrate the case regardless of cultural context.

On the other hand, formal mathematics also acknowledges the idea of plural certainty. There are the cases where mathematicians cannot reach a common agreement, so a problem cannot ultimately be resolved. The only resolution is to recognize a plurality of systems and how these systems are chosen based simply on human preferences. Because of this pluralism, mathematicians cannot agree on the philosophical basis of mathematics. In this view, mathematics is much like a house without a foundation. Mathematics floats, so it is more appropriate to speak of mathematical systems as houseboats that shift positions on the waves of reality.

In the real world, mathematics comprises univocal and plural arguments and formulas. When there is agreement (univocal) on mathematical certainty, it is usually based on a consistency within a certain formula or system. Many schools of thought in mathematics gather around these certainties. However, lack of agreement arises because of the problems of incompleteness and nondecidability of a given formula or system. That formula or system must remain open and cannot be considered an absolute certainty.

In the twentieth century, our recognition that formal mathematics comprises both univocal and plural realities had an impact on our view of metaphysics as well. Metaphysics traditionally tried to find absolute certainty, conveying that in a symbolic language. The theorems of incompleteness and nondecidability can lead us to think that, in general, making absolute statements of any type is not possible.

Nevertheless, beyond the scope of mathematics, every human being needs to ask ultimate questions about existence in general and about his own existence in particular. The parallel to this in mathematics is the principle of consistency (technically called the principle of noncontradiction), which is the widest basis for certainty and truth in mathematics and in life. Mathematics cannot cease to be consistent. Mathematics exists because it is consistent. Therefore, consistency is a logical principle that unites metaphysics and mathematics.

By definition, consistency must exist in mathematics and meta-physics. The attempt to negate or disprove a principle, for example, must be carried out with a consistent language since, if this were not so, the negation would cease to be logical. No matter which way we turn, consistency is required as the most basic logical principle in human thought and communication. Human knowledge of the laws of nature may vary, but the laws of nature—by definition—cannot cease to be consistent.

As we begin our survey of the formalization of mathematical language in the twentieth century, we adopt two kinds of overall perspectives. One is that study of the foundations of mathematics has shown that mathematics is plural. The other is that the consistency of thought required in mathematics is where we have found the absolute certainties to reside. With both pluralism and univocal certainty (consistency), we can conclude that mathematics and science remain open to—and often friendly to—the language and ideas of metaphysics.

To understand how the problems of incompleteness and non-decidability arose in the modern history of mathematics, we can survey three stages in the process. First was the formalization of mathematical language. Second was the realization that mathematical systems are incomplete and plural. Third were the findings that, even with the powers of the algorithm, some mathematical calculations cannot be decided. These milestones and realizations introduce us to many of the greatest names and personalities in modern mathematics.

First Stage: Formal Mathematics

One of the first steps that helped the emergence of formal mathematics was the appearance of the non-Euclidean geometries, showing that geometry did not depend on the traditional intuitions that supported it for so many centuries.

Euclid proved deductively in the *Elements* the fundamental

propositions of geometry based on certain axioms. An axiom is a mathematical statement that is accepted as evidence without proof. Euclid's fifth postulate (or axiom) states that if a line segment intersects two straight lines forming two interior angles on the same side and these add up to less than two right angles, then the two lines, if extended indefinitely, meet on that side on which the angles add up to less than two right angles.

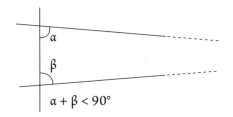

$$\alpha + \beta < 90°$$

By describing lines in infinity, the fifth postulate differs from others in the *Elements*. The fifth postulate operates on the intuition of the infinite, which separates it from the intuition about finite objects. This contrast prompted some mathematicians to try to avoid using Euclid's reference to infinity, since it seemed sounder to refer only to finite objects. For example, Proclus of Greece (d. 485 CE) wrote a commentary to Euclid in which he attempted to derive the fifth postulate from the other postulates, which do not include the concept of the infinite.

The problem of whether the fifth postulate of Euclid is an axiom that is independent of the others—that is, whether it is possible or not to deduce it from the other postulates—intrigued geometricians. Among those who studied this problem were the Italian Jesuit Girolamo Saccheri (d. 1733) and the Swiss-German philosopher Johann Heinrich Lambert (d. 1777). In his work *Euclides ab Omni Naevo Vindicatus* (*Euclid Preserved from Any Error*), Saccheri tried to prove the fifth postulate of Euclid by supposing first that it was false. In other words, he tried to derive a contradiction from the other postulates and thus negate the fifth postulate, but he found none. If he had found a contradiction, he would have proved that the fifth postulate is deduced from other postulates and thus does not stand on its own.

The Russian mathematician Nikolai Ivanovich Lobachevsky (d. 1856) followed the same route as Saccheri and Lambert. He assumed, contrary to the fifth postulate of Euclid, that "through a point which is exterior to a straight line, it is possible to draw at least two lines parallel to this straight line." By adopting a postulate contrary to Euclid's fifth postulate, Lobachevsky developed a new geometry that did not produce any contradictions. However, Lobachevsky had to call his new geometry "imaginary" because it did not match any real model in the world.

In his reflections on non-Euclidean geometry, Lobachevsky relativized the spatial intuition that gave traditional geometry its explanatory power. The force of Lobachevsky's non-Euclidean argument was based only on the absence of contradiction in his deductions. But at the same time, he could not prove that non-Euclidean geometry is consistent. He showed only the absence of inconsistency in his theorems. In the end, therefore, Lobachevsky argued for the existence of two different geometries: the traditional Euclidean geometry and the new non-Euclidean geometry.

The consistency problem left over by Lobachevsky was attacked in 1871 by the German mathematician Felix Klein (d. 1925). He proved that, if classical Euclidean geometry is consistent—that is, if contradictions cannot be deduced from its axioms—then non-Euclidean geometry is also consistent. In this way, Klein reduced the problem of proving the consistency of non-Euclidean geometry to the problem of proving the consistency of classical Euclidean geometry. However, the problem of the consistency of classical Euclidean geometry was still open. There was no proof that classical Euclidean geometry is consistent.

David Hilbert (d. 1943) presented in 1899 another attempted solution to the consistency problem, publishing his findings in *Grundlagen der Geometrie* (*Foundations of Geometry*). In this work, he proved that Euclidean geometry is consistent under the supposition that the theory of real numbers was consistent. Hilbert reduced the problem of proving the consistency of Euclidean geometry to proving the consistency of the theory of real numbers. Still, at the opening of the twentieth century, the consistency of Euclidean and non-Euclidean geometry—based on the consistency of real numbers—remained an open problem for mathematics to address.

A kind of breakthrough came when empirical science gained its first hints that non-Euclidean space—a curved kind of space bent by gravity—was indeed a physical reality. As we recall, Lobachevsky considered non-Euclidean geometry to be imaginary, but now it was more than a fantasy. Non-Euclidean geometry was a system that helped explain the physical reality of a curved-space universe as detected by experiments that were trying to test Einstein's relativistic theory of gravity (that on the largest scales, gravity bends a beam of light). Today, non-Euclidean geometry is considered a valid theory and an empirical fact. (See Figure 6.1.)

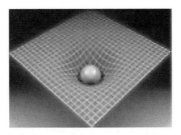

FIGURE 6.1

Reprinted by permission from Macmillan Publishers Ltd. from J. Horgan, "Physicists Plan Search for the Known Unknowns," *Nature* (February 1, 2007), 468–69.

As a by-product, moreover, this struggle to deal with non-Euclidean geometry forced modern mathematics to question deeply the consistency of all mathematical theories. To attack this problem, mathematics needed a formal language each of the systems could use to test their consistencies. Hence, the project to formalize all mathematical language was born.

This attempt to create a universal language for all mathematics was a bold move, but this quest has always been at the heart of science, which is in search of universal laws for all of nature. The search goes back through the ages. At the start of the fourteenth century, for example, a philosopher from Majorca, Spain, named Ramón Llull tried in his *Ars generalis ultima* (*The Ultimate General Art*) to find a universal symbolic language for all knowledge— a dreamy ambition, of course. In reality, we had to wait until the nineteenth century before a universal sign language was available for even basic logic and algebra. This was the start of the complete formalization of mathematical language, which stretched into the twentieth century and involved several great names in mathematics, as we shall see below. A convenient place to begin is in the early 1800s with the English mathematician George Boole.

Born to a modest family in Lincoln, Boole was largely self-taught in mathematics. Nevertheless, in 1854, at the age of thirty-nine, he published *An Investigation into the Laws of Thought*. In this work, Boole described with algebraic methods what we now call propositional logic. This was a first step in the formalization of mathematical language. In this methodology, Boole outlined the kinds of concise propositions that could be built from other simpler propositions, using connectives like "and," "or," and "not." His work did not, however, analyze the functions and relations that might be present in these propositions in a wider language.

That was a language achieved by the German mathematician Gottlob Frege (d. 1925). After studying mathematics, philosophy, and physics at the universities of Jena and Gotinga, Frege went on to develop a system of formal logic that we call first-order predicate logic. This system uses the reasoning of arithmetic and geometry, but it also determines whether all the elements (or some elements) in a given domain of discourse have certain mathematical properties.

Frege published his system as *Concept Notation* in 1879, adding the subtitle, "A Language of Formulas for Pure Thought in the Likeness of Arithmetic." In this project, he presented mathematics as a superstructure that rose on the foundation of formal logic. He introduced specific formal signs to denote functions and relations. For example, he introduced quantifiers such as ∀ and ∃, which refer, respectively, to all and to some elements of the domain of discourse. In effect, *Concept Notation* made it possible to insert in mathematical statements the exact rules of logical inference by which to conduct formal mechanical operations. These rule-based operations were designated simply by the way the formal signs were placed in the mathematical language.

A contemporary of Frege was the Russian-born mathematician Georg Cantor, the son of a wealthy Danish businessman and his Russian wife. The family eventually moved to Germany, where the climate was better suited to the father's health. Though a lover of music, the young Cantor pursued the study of mathematics in Switzerland as well as Germany and eventually became a professor at the University of Halle.

From that German outpost, he made his major contribution to the formalization of mathematical language. Cantor did this with his theory of sets. If Frege had created a syntactic language for all mathematics, Cantor organized and defined the actual objects of mathematics through set theory. Cantor defined a set as "a collection into a whole of definite and separate objects of our intuition or our thought." By presenting this definition, Cantor proposed a uniform way to compare and relate all possible mathematical objects. Each object was a collection of elements. This gave mathematics a standard semantics so that it could deal in the same way with numbers, points on a map, and the seconds in the day. Each was a set.

As we can see already, the process of formalizing modern mathematical language was, on one hand, a building up of certain elements, but on the other hand, mathematicians showing each other the contradictions of their work. Through this process, the best kinds of languages and the worst contradictions were jettisoned and something like a workable formal language was reached, even though it was finally a plural system. (In Appendix 6, I have placed the story of how the English mathematician Bertrand Russell, for example, found a paradox in the system of Frege just as Frege was publishing his master works.) Overall, this historical process led to the ability of mathematics to use its own methods to study itself.

On the foundation of those who came before him, the German mathematician David Hilbert put forward the idea of metamathematics, which is the use of mathematical language and principles to analyze mathematical systems themselves as a whole ("meta" meaning beyond or above). Hilbert began his mathematical studies in the city of Königsberg, where his father had been sent to serve as a judge in the court system. Hilbert also attended courses in Leipzig and in Paris, where he was exposed to some of the greatest mathematicians of his day, including Henri Poincaré, Camille Jordan, and Charles Hermite.

Hilbert began to teach mathematics at the University of Göttingen in 1895, and his most significant contribution to the formalization of mathematics materialized in a course that he taught three years later. The course's title was "Elements of Euclidean Geometry." In this, Hilbert showed how geometrical theorems can be deduced from axioms by pure logic, bypassing the intuition that early geometry had assumed. By using pure logic in all geometry, he showed the logical-formal value of mathematics. As Hilbert once said, theorems must be valid even if, instead of being points, lines, and planes, the objects being looked at are "tables, chairs, and glasses of beer." In other words, all of the objects must obey the axioms (since the mere intuitions of geometric shapes are no longer there to rely upon).

So by the time of Hilbert's illustrious career, the idea of metamathematics was in full bloom. Now it was assumed that mathematics as language should be able to speak about mathematics itself as the object of that language. Mathematical objects were represented by mathematical formulas, reasoned from mathematics itself. Mathematical theories were ruled true or false by relating them to mathematical models.

The goal of the metamathematical program was very ambitious. It hoped to find a final and universal formal language. With such language, perhaps mathematicians could finally prove that mathematical theories are consistent (with no contradictions), complete (all formulas follow the axioms), and semantically decidable (having a procedure to decide the truth of a theory). As we see in the next section, this ambition of formal mathematical language was being undermined at the same time that it was trying to build itself up.

SECOND STAGE: THE INCOMPLETENESS OF ARITHMETIC

The undermining began with new findings in arithmetic. This is the story of Kurt Gödel, who was born into a German family in the city of Brno (now in the Czech Republic). After studying locally, he attended the University of Vienna, where he was very impressed by a lecture given by Hilbert on completeness and consistency in mathematical systems. At age twenty-three, Gödel wrote his doctoral thesis on the completeness of first-order predicate logic.

In his thesis, Gödel proved that all the propositions which are true in all the models of first-order predicate logic can be obtained from a system of axioms. This is called "completeness of the logic of predicates of first order." Gödel naturally found this very satisfying, and it whetted his appetite to seek the same completeness in other kinds of mathematical systems. So he next turned his attention to arithmetic. However, when he applied the same analytical methods of metamathematics, he ended up proving that the formal system of arithmetic is incomplete.

This incompleteness of arithmetic can be stated in the following way. Despite the fact that we might add additional first-order axioms to the system of arithmetic, and that the new axioms do not lead to a contradiction (that is, a proposition of the type A and not A), there can still be a proposition U that will be nondecidable in the system of arithmetic. That is, within the formal system of arithmetic, U cannot be deduced from the axioms. Therefore, it is not possible "to decide" whether U belongs to the system or not. Gödel also proved that if arithmetic is consistent, then it is not possible to prove within the system of arithmetic the formal proposition that expresses the consistency of arithmetic.

In discussing Gödel's dramatic finding, mathematicians should not confuse the theorem of the *completeness* of first-order logic with the problem of the *incompleteness* of arithmetic. In the first case, the completeness of the calculation of first-order logic is affirmed. In the second case, the incompleteness of arithmetic is affirmed. Gödel's conclusion resounds across modern mathematics: arithmetic is an incomplete system, which in turn proves that we cannot construct an axiomatic system from which we can deduce all the propositions valid in arithmetic. From any axiomatic system, we can only deduce part of the arithmetic.

We can look at the implications more closely by creating an analogy of going to the store to buy something. On such a trip on Monday, we choose the items we wish to buy and determine the price by using arithmetic. On Wednesday at the store, our purchase is more elaborate, and the arithmetical calculations more complex, but we resolve this as well. Now think of the arithmetic we used on Monday as if we were in a dark room, and a nearby lantern is what allowed us to find the price of that purchase. On Wednesday, a lantern in a different corner of the store allowed us to decide that price. For mathematics to be complete, we would need the entire room lighted by a single lantern. However, now that we know arithmetic is incomplete, we must proceed as if a lantern is needed to light our way here and there. Incompleteness tells us that it is not possible to illuminate everything through an axiomatic system. Any illumination we make will be partial.

Nevertheless, the idea of incompleteness does not destroy our assumption that we are being consistent in our thought. Consistency led Gödel to his conclusion. He said that if arithmetic is consistent, then it cannot be complete. At the same time, we cannot prove the consistency of arithmetic from within the arithmetic. We must go outside of the arithmetical system and use metatheory, which we presuppose to operate in a consistent way as well.

Gödel's findings introduced a crisis to Hilbert's metamathematical program. As a result, all of mathematics had to reassess itself. Was it possible to have a single mathematics, or was the project destined to come up with a plurality of languages and mathematical systems?

Hilbert and Gödel based their research and conclusion on the premises of classical logic and mathematics. In that tradition, a proposition can be proved true if we can contradict its negation. Furthermore, classical mathematics accepts the existence of infinite sets. These ground rules guided Hilbert to seek a universal system, but also led Gödel to conclude that some component systems of mathematics (such as arithmetic) cannot be complete. Classical mathematics itself led us to this impasse.

To solve the impasse, other schools of mathematics arose. They were based on rejecting some of the basic presumptions of the classic tradition, which explains the origins of the intuitionist/constructivist school of mathematics. This group relies on stricter assumptions about the basis of mathematics and accepts only those propositions that are directly proved as valid. Depending on the nonclassical school, the existence of sets with infinite elements is also only narrowly accepted, though usually rejected.

The Dutch mathematician L. E. J. Brouwer (d. 1966) was an outstanding leader of the constructivist school of thought. Gödel's proofs of incompleteness showed that classical arithmetic had severe limits, so Brouwer offered a new version of mathematical reasoning. His was a narrower version since it only allowed mathematical objects that were directly proved (which excludes those based on contradiction of a negation). Fortunately, the classical and constructivist schools of mathematics can coexist, but separately. Their main link is the fact that constructivist mathematics at least accepts the direct proofs of classical mathematics.

This existence of separate systems based on different conceptions of logic is not contradictory. In other words, using a meta-language based on classical logic, we can at least understand and reflect on constructivist logic. From the outside, we can see that both logics have different approaches, but neither of them proves the falsity of a theorem of the other. Moreover, we can consider constructive mathematics as a subset (the constructive part) of classical mathematics.

The plural coexistence of the systems is also possible because both Hilbert and Gödel described mathematical objects based on Cantor's set theory. In other words, the use of sets is common ground in classical and constructivist systems. Using sets, we can say that a mathematical proposition is true in a certain set if the components of this set fulfill the relations that appear in the proposition. If they are not fulfilled, we say that the proposition is false. For example, we say that the proposition "2 + 3 = 5" is true in the set of natural numbers because the addition function is interpreted in the set of natural numbers and is applied to the sets that we designate as "2" and "3" and provides the set we designate as the number "5."

But set theory is not the only possible description of the mathematical objects. Subsequently, other theories have developed that begin with abstract and general descriptions of mathematical objects other than sets. Thus, a French group of mathematicians (who remained anonymous by publishing under the fictitious name "Nicolas Bourbaki") attempted to construct all modern mathematics based on the concept of structure. Later, the concept of category appeared as a basic description of mathematical objects. Categories are not based on the concepts of sets and belonging, but on the concepts of function and composition.

In conclusion, while the classical and constructivist systems can coexist and share some common elements, as formalist systems they are incompatible approaches. Nevertheless, we can compare them by going outside of each system and using the methods of meta-language, which looks for consistency. The comparative metastudy of incompatible systems makes it possible to isolate each system while interrelating the systems from an outside viewpoint.

This plurality of mathematical systems is like a cluster of grapes; each system (like a single grape) has internal consistency. We cannot unite two systems, making one system from two. However, we can unite them in a cluster. The system of twigs holding the cluster of grapes together is like the mathematical metalanguage that looks at the whole bunch from the outside. The metalanguage allows us to have a vision of the plurality of formal systems that is simultaneously open and consistent.

THIRD STAGE: THE FORMALIZATION OF THE PROCESSES

In the preceding stories of Hilbert, Gödel, and Brouwer, we have seen how the incompleteness of arithmetic has undermined the mechanistic ideal of a single formal language and prompted the creation of plural systems. Next we see how Gödel's findings, plus our new knowledge of the mechanistic algorithm (that is, a mechanical procedure of calculation), has revealed that some problems in mathematics simply cannot be decided. This is another blow to the ideal of a single universal mathematical system.

The decision problem, known in German as *Entscheidungsprob-lem*, arises when we are faced with a set of infinite elements and ask whether a given element belongs to the set or not. Given the problem of incompleteness, we can at best make conjectures. One example is the conjectural statement that there are no perfect odd numbers. A number is perfect if it is greater than 1 and is equal to the sum of its divisors other than itself. For instance, the first four perfect numbers are 6, 28, 496, 8128:

$6 = 1 + 2 + 3,$

$28 = 1 + 2 + 4 + 7 + 14,$

$496 = 1 + 2 + 4 + 8 + 16 + 31 + 62 + 124 + 248$

$8128 = 1 + 2 + 4 + 8 + 16 + 32 + 64 + 127 + 254 + 508$
$+ 1016 + 2032 + 4064$

In the history of mathematics, several conjectures have been made on the perfect numbers that later turned out to be false. However, one conjecture that has not been proved either way is that all perfect numbers are even numbers. Another case is the famous conjecture of Christian Goldbach, the Prussian mathematician. In 1742, Goldbach proposed the following: every even number greater than 2 can be written as the sum of two prime numbers. It is easy to check that this conjecture is verified for the first even numbers greater than 2:

$4 = 2 + 2; 6 = 3 + 3; 8 = 3 + 5; 10 = 3 + 7; 12 = 5 + 7 \ldots$

and arguments of a statistical nature make it possible to affirm that, as regards very large numbers, this conjecture is very probable. However, it has not been possible to provide a rigorous proof that the conjecture is true for any even number however big it may be. The conjectures of Goldbach and the conjecture that the perfect numbers are even are two examples of arithmetic propositions where it has not been possible to decide whether they are true or false.

After Gödel, that mathematics has this feature of unresolved conjectures is not surprising. By using a metamathematical theorem, he showed that there was no complete system of axioms on which to build arithmetic. However, one thing that Gödel did not do is translate his theorem into a mechanical form of calculation, the algorithm language. This kind of language can be formalized and understood by a machine, which today, of course, means a computer. In other words, nobody so far—not Boole, Frege, Russell, Hilbert, or Gödel—had taken their formal languages and described how they can be used in the automatic calculation of computation.

The idea of a mechanical process is inherent in the uses of informal calculations that were already understood in ancient times. In early logic and algebra, an informal kind of algorithmic process was employed. What was lacking until the twentieth century was a formal mathematical language to write down the instructions for such a calculation. These instructions create a formal mechanical process, which has four properties:

1. It is governed by a finite number of precise instructions. In computers, this set of instructions is called a program.
2. The instructions can be executed in a finite number of steps. In computers, we say the process must terminate.
3. The calculator (machine) is not allowed to take initiative. The instructions are executed mechanically.
4. A human being who had sufficient time could also execute these instructions using a pencil and paper, which means the execution can be represented in the language of formal mathematics.

With such a mechanical procedure, logicians and philosophers had long dreamed of a machine that could calculate all logical deductions. This is also called the effective method, and it was in the years before World War II that a computing machine was conceived of to carry out this method based on a set of instructions. Those instructions—the formalization of the algorithmic processes—were the brainchild of two mathematicians at about the same time, Alan Turing (d. 1952) in Britain and Alonzo Church (d. 1995) in the United States.

By his achievements, Turing is considered one of the fathers of theoretical computing. He spent much of his English childhood separated from his father, who was in the British civil service in India. He studied at Cambridge University and made history in 1936 with his paper on how to program and build a so-called Turing machine, a computing program and device that could represent any algorithm. In other words, Turing's paper provided a formalized mathematical language for the effective method.

Turing was not the only one making such progress. At the same time, Alonzo Church, a brilliant student at Princeton University (who later taught at Princeton and the University of California–Los Angeles), was also pioneering a formalized algorithm. Indeed, a few months before Turing's 1936 paper, Church had presented a somewhat different formalization of the algorithms. Church's formalization is called the lambda calculus. From different angles, the Turing machine and the lambda calculus described essentially the same set of algorithmic procedures.

The Church-Turing accomplishment was a historical milestone on par with Frege's formalization of first-order logic. Just as Frege's first-order logic gave us a standard formal language for arithmetic, Church and Turing gave us a standard formal description of an algorithmic process that a machine can execute. But as we shall see, even a perfectly designed machine may not be able to overcome the logical problem of undecidability.

The big puzzle for the Turing machine was whether after a certain period it would stop, representing resolution of a complete system of calculation, or whether it would keep running in an infinite cycle. In fact, the latter was the result that Turing obtained. The mechanism of the Turing machine confirmed that it is not possible to formally prove, in all cases, whether a given program will stop and give an output. The Turing machine confirmed mathematical undecidability, strengthening Gödel's theorem of incompleteness.

The image of a Turing machine and a formalized language of instructions joining forces to master a universal mathematics is a compelling one. But the two finally could not join in this conquest. The lesson is that a physical discourse and a mathematical discourse are not exactly equal partners in our understanding of logic and nature. The physical discourse must take the mathematical discourse into account. However, the mathematical discourse does not have to take the physical discourse into account. It has been proved that mathematical discourse is undecidable, although this does not necessarily demonstrate that physical discourse is undecidable.

To put it differently, the theorems of undecidability demonstrate that we cannot deduce all of mathematics from a single set of formal axioms. Importantly, this conclusion also suggests that we do not need to use all mathematical theories when we try to explain physical theories. For example, non-Euclidean geometry can be used to explain the relativistic theory of gravity, but the theorems of arithmetic do not have to speak about how gravity acts. Because all of mathematics is not a unity, it is sufficient to use only a limited number of mathematical propositions to explain a given fact of nature.

As another example, we do not need to know whether Goldbach's conjecture is true in order to know mathematically the conduct of elementary particles. We could have a complete theory of the interaction of elementary particles based on a finite set of axioms that could be interpreted in a single formal model of the world.

Thanks to Church and Turing, computing has decisively affected the development of the applied sciences. For the first time, formal models of mathematics are taking on the applied dimension of pragmatism, experimentation, implementation, and efficiency. To do this, computer scientists have drawn upon a plurality of logics depending on the output sought. This plurality of logics opens new perspectives for scientific language as each of these reflects a partial dimension of reasoning. In addition, while applying this partial formalism, computing is also discovering new probabilistic properties in algorithms.

Finally, pluralism and undecidability have forced us to recognize the growing complexity of mathematical logic and systems. More and more we are trying to find different solutions to the same problem. The challenge today is to discover criteria to determine the best solutions. The best is always the simplest, but in the application of mathematical formalisms, we are still grappling with a growing complexity of computer programs as we try to process the data of the empirical sciences.

Mathematics, Reality, and Metaphysics

One other way to think of incompleteness and undecidability is the fact that the human mind is creating many of our mathematical systems ad hoc. Mathematics is not outside the mind, as Platonism and the classical tradition have held. We are next forced to ask whether, if this is true, mathematics can give us a true image of the universe, or is mathematics a kind of peephole or warped mirror that gives us a limited, or even distorted, view of reality? In our era, the three big questions about mathematics and reality are as follows: Does math tell us about empirical reality? Can chaotic and probabilistic systems in nature ever be intelligible through mathematics? How do we further clarify the languages of signs and symbols?

Throughout the twentieth century, we have generally adopted a naïve scientific realism that says the knower can simply know the real object. That view, often called positivism or materialism, is now in crisis. In the case of quantum physics, we know that the empirical observations are not totally objective, because our observation interferes with nature in the process. Nevertheless, when these quantum theories are expressed in mathematics, we seem to believe that our observation is objective. In other words, we rely on mathematics to persuade ourselves that we are reading nature objectively.

Before I go into the problematic side of this reliance, we must acknowledge that mathematics has indeed increased our objective grasp of reality. It can formulate lawlike principles. Mathematics allows us to make predictions, which are the very basis of scientific research and testing. The same consistency in mathematics produces scientific instruments of exquisite sensitivity and accuracy.

Nevertheless, not only do mathematical systems have limits, but objective reality itself is reluctant to allow itself to be controlled. Reality resists our mathematical attempts to enclose it within determinist orders. In chaotic and probabilistic orders of nature, *unpredictability* appears together with predictability. The idea of an attractor in a chaotic system illustrates this enigmatic mixture. In a chaotic system, its elements can move to an attractor from different points. The attractor's structure can be deterministic (a normal attractor) or it can be probabilistic (a strange attractor), but in either case, the process reveals order emerging from disorder, something that mathematics is not yet equipped to understand entirely.

In chaotic systems as well, we can observe that a very small variation at some location can substantially alter the trajectory of the entire system, causing it to converge on one kind of attractor or another. Observing this kind of bizarre phenomenon, we naturally wonder about the intelligibility of the world.

Another kind of chaotic system is a probabilistic order. In this system, physical events occur independently of each other. Each event is unpredictable in relation to the others. Nevertheless, a probabilistic pattern to chaos can be captured in mathematical laws such as statistics. Mathematics itself has a kind of randomness and probability as well. By the use of statistics, for example, we can forecast the average distribution of prime numbers even though the position of each prime number seems to be random across the numerical system.

In summary, the chaotic and probabilistic orders have transformed our view of the role of rationality in our empirical study of the world. Like Galileo, we can say that nature is written in a mathematical language. However, unlike Galileo, we can no longer presume to reduce mathematical language to triangles, circumferences, and other geometric figures. The scientific axioms that we had relied on until the twentieth century have now lost much of their power of explanation. To know and explain nature today, we need to use signs and symbols.

The formalization of mathematical language has given us those signs. Today, the signs can be handled by computers to approximate certain kinds of human thought. We are not exploring whether a machine can think completely like a human being. Computers have shown that they can carry out any mathematical thinking based on signs. Within the limits imposed by the formal languages, a machine can produce propositions and systems of thought, and, in some cases, with more capacity than the human mind. A computer can develop its own programs. But these programs will always be written in a language of sign. What a computer cannot do, however, is produce symbolic thought, which is a unique feature of the human mind and the language of metaphysics.

In a famous lecture that Turing gave on February 20, 1947, at the London Mathematical Society, he addressed the topic of whether a computer can simulate human thought. He rephrased the question as whether a computer could be programmed to learn in such a way that it imitates the human trait of learning by committing errors:

> I would say that fair play must be given to the machine. Instead of it giving no answer we could arrange that it gives occasional wrong answers. But the human mathematician would likewise make blunders when trying out new techniques. . . . In other words then, if a machine is expected to be infallible, it cannot also be intelligent. There are several mathematical theorems which say almost exactly that. But these theorems say nothing about how much intelligence may be displayed if a machine makes no pretence at infallibility.

Given that human beings cannot deduce all the formal propositions from a system of axioms, we must try out several axioms and discard the failures. Therefore, Turing had to conclude that computers could not be more infallible than human beings.

At this point Turing pointed to the game of chess as a kind of thought experiment about the mental ability of humans and computers. For both humans and computers, there must be an element of risk in a game such as chess, even though chess—which has rules and a limited number of pieces and possible moves by each piece—is ultimately a theoretically decidable game. Nevertheless, in chess, humans and computers must take the risk of making certain propositions that cannot be deduced from other accepted propositions.

The computer must make risky decisions for the simple fact that even in finite chess, the number of possible scenarios is large enough that no computer could possibly learn all possible strategies for victory. Nevertheless, computers still have the power to make decisions guided by heuristic programs, that is, programs that follow rules of thumb. In contrast, human players make risky decisions based on intuition. This difference is what fascinated Turing: he compared heuristic programs with human intuition, both of which had knowledge of the moves allowed to each chess piece.

In the last fifty years, however, Turing's thought experiment has changed. Better computer programs and heuristics—beyond Turing's imagination, perhaps—allow computers to beat chess champions. Still, the comparison between computer heuristics and human intuition is interesting. This is a human-computer contest over mastery of formal signs, and because chess is indeed a theoretically decidable system, the computer is finally able to manipulate signs more powerfully than the human mind.

Unlike the game of chess, however, the realm of all mathematics is not theoretically decidable, as we noted above, and therefore the fact of error remains unavoidable in the human use of mathematics. Error is a common concern in the normal perception of our senses, as in seeing a mirage on a desert horizon, and we know that even our best measuring devices have margins of error. This knowledge helps us recognize that formal mathematical thought is also open to errors. Mathematical results cannot be complete, which has led us to use many possible systems, none of which is predetermined in all cases. We also see the world as made up of probabilistic and chaotic natures. On all fronts, mathematics is faced with indeterminacy, and it must constantly choose between several possibilities.

If mathematics has led the evolution of human rationality across history, then where does our notion of rationality stand today? Whatever that definition is will also be reflected in our metaphysical outlooks. During the Enlightenment of the seventeenth century, a mathematician such as Leibniz had a deterministic conception of rationality with its doctrine of a sufficient reason for everything. Consequently, Leibniz also had a deterministic vision of the metaphysical questions: he viewed God as controlling absolutely every action in the universe.

We do not presently have a deterministic vision of rationality, which naturally influences our metaphysical questions. Leibniz proposed that, in order to terminate conflicts, when two persons are involved in litigation, they should define the concepts within a formal system of calculation. The parties involved in discussion should simply sit down and make calculations. Leibniz thought that all conflicts could be resolved in this way. Nowadays, we see Leibniz's proposal as naïve. We cannot even perform such calculations within a unique axiomatic system.

The opening of our rational world to risk is not an opening up to irrationality. We are still approaching these metaphysical questions with a consistency that characterizes rational thought. By our rational assessment, we look for the consistency—that is, the lack of contradiction—within each of the plural systems. We can also judge that these systems can coexist because they do not mutually exclude one another.

To keep rationality in our world, we are required, in the end, to make a metaphysical leap. The formal language of mathematics cannot prove the consistency. Rather, our belief in consistency is a metaphysical presumption. Hence, consistency becomes a rational and scientific value, but also a theological value. Theology must seek to be as consistent as possible with logic, math, and science in a world that is open to incompleteness, undecidability, risk, and error. This lesson may be the one that the history of mathematics offers for the modern dialogue between science and religion.

CHAPTER 7
Propositional Logic

NOW THAT WE HAVE established a plurality of mathematics and recognized its limits, we can nevertheless learn how the formal language of classical mathematics works. In this chapter we look at the most basic kind of language: propositional logic. As we saw, the mathematician George Boole gave us the earliest system of propositional logic. Later, the mathematician Gottlob Frege developed first-order predicate logic. We will speak of these two languages as L0 and L1, respectively. (To simplify our discussion, I have added appendices that look at L0 and L1 in much more technical detail.)

Together, propositional logic and first-order predicate logic make up the nucleus of classical mathematics. They show us why mathematics continues to have a privileged role in our knowledge of the world, especially in science. In fact, the usefulness of formal logic derives from its similarities to our normal human languages. Natural languages rely on words and symbols to convey structure and meaning, so they cannot be reduced to a formula of signs as seen in formal logic. Nevertheless, parallels exist in the way we can approach natural and formal languages.

When we learn a natural language at school, for example, we usually begin by looking at how words operate in sentences. The same is approximately true in the language of formal logic. There are three linguistic dimensions to formal logic. The first is the syntax, how the language is organized. Second is the semantics, which looks at the meanings conveyed by the formal language. Finally there is pragmatics, which studies the particular ways that we can use logic.

Furthermore, all language is complicated by the inherent fact that we must use a language to talk about a language. In this book, for example, we are using the English language to talk about the language of formal logic. Observing this kind of necessity, mathematicians have offered a distinction between a metalanguage and an object language. In the example above, English is the metalanguage while formal logic is the object language. When we use Spanish to talk about English, the first is our metalanguage and the second our object language. This distinction may seem odd, but it becomes necessary when a single type of language becomes the topic of our analysis. Such is the case when we use English to talk about English, mathematics to talk about mathematics, or formal logic to talk about formal logic.

Now back to the declarative proposition that characterizes L0. The heart of formal logic is the declarative proposition. It is the first level of logic. Declarative propositions are building blocks in the language of propositional logic. (First-order logic goes deeper by analyzing the internal composition of these declarative propositions.) What are some examples of a declarative proposition? Any such proposition affirms or denies a fact as true. For example, the following are declarative propositions:

Luisa is my friend.
3 is greater than 4.
London is in Europe.

However, propositions such as "Go away!" or "How old are you?" are not declarative; they are a command and a question, respectively. They do not affirm or deny a fact. Therefore, these statements cannot be declared true or false. As we shall see, logical argumentation is based exclusively on a series of declarative propositions.

As noted above, a proposition affirms a fact and can therefore be true or false. But there are several kinds of propositions. The most basic is the atomic proposition, which means that the statement is a single proposition; it is not made up of more elemental propositions. The formal language we call L0 is so basic because it is built entirely from atomic (as in atomistic) propositions.

We designate atomic propositions through letters: p, q, r, etc. For example, "John loves Luisa" and "2 is an integer number" are atomic propositions. We write p ≡ "John loves Luisa" in order to indicate that the atomic propositional formula p represents the atomic proposition John loves Luisa. (We also use the logical constants ⊤, ⊥ as atomic propositional formulas that represent a true or false proposition, respectively.)

Another kind of proposition is a compound proposition. This is constructed through connectives that can be monadic or binary. (In our discussion below, we use a monadic connective that we call negation, and we also use three binary connectives that we call conjunction, inclusive disjunction, and conditional.)

First, we look at negation. A proposition may be the negation of another proposition. For example, "John does not love Luisa" and "2 is not an integer number," are the negations of "John loves Luisa" and "2 is an integer number." We formally designate the negation connective by the sign ¬, which we place before a propositional formula. For example, if p ≡ John loves Luisa, then ¬p ≡ John does not love Luisa.

We use Greek letters φ, ψ, χ, etc. in order to represent all kinds of propositional formulas whether they are atomic or compound. Another important designation is to represent the semantic status of propositions as either true or false, which are represented by the signs 1 and 0, respectively. The following semantic table of the symbol ¬ indicates that a propositional formula φ is true if its negation ¬φ is false and vice versa:

φ	$\neg\varphi$
0	1
1	0

Now we look at the three binary connectives, beginning with the conjunctive type. In this case, the propositional formula $(\varphi_1 \wedge \varphi_2)$ is the conjunction of two propositional formulas φ_1, φ_2. The conjunction of φ_1 and φ_2 will be true if and only if φ_1 and φ_2 are both true. In English there are several conjunctive particles with which we can construct a true statement from two true statements. For example, if p ≡ Mary is shrewd and q ≡ Peter is ingenious, then the conjunction $(p \wedge q)$ can be expressed in English with any of the following statements:

Mary is shrewd and Peter is ingenious ≡ $(p \wedge q)$
Mary is shrewd but Peter is ingenious ≡ $(p \wedge q)$
Mary is shrewd however Peter is ingenious ≡ $(p \wedge q)$
Mary is shrewd although Peter is ingenious ≡ $(p \wedge q)$
Mary is shrewd even though Peter is ingenious ≡ $(p \wedge q)$
Mary is shrewd while Peter is ingenious ≡ $(p \wedge q)$, etc.

In all these cases the formula $(p \wedge q)$ will be true if both p and q are true. We can express this fact in all these cases through the semantic table for any two formulas φ_1, φ_2:

φ_1	φ_2	$(\varphi_1 \wedge \varphi_2)$
0	0	0
0	1	0
1	0	0
1	1	1

The second kind of binary connective is the inclusive disjunction. In this case, the propositional formula $(\varphi_1 \vee \varphi_2)$ is the inclusive disjunction of two propositional formulas φ_1, φ_2 connected through the logical sign "$\vee$." The inclusive disjunction of two formulas is true on condition that at least one of the two is true. The semantic table of the inclusive disjunction is as follows:

φ_1	φ_2	$(\varphi_1 \vee \varphi_2)$
0	0	0
0	1	1
1	0	1
1	1	1

In English the particle "or" may mean both inclusive and exclusive disjunction. The exclusive disjunction of two propositions is true only in the case that one of the two propositions is true. For example, if a restaurant menu offers fruit or cheese as dessert, it would be right to ask if "or" is inclusive or exclusive—that is, can we take both or not?

The third kind of binary connective is conditional. In this case, the propositional formula $(\varphi_1 \rightarrow \varphi_2)$ is the conditional of two propositional formulas φ_1, φ_2. In English there are several ways to express this construction. For example, if p ≡ Mary is shrewd and q ≡ Peter is ingenious, then $(p \rightarrow q)$ represents any of the following statements:

if p, then q;

p implies q;

p only if q;

p is a sufficient condition for q;

q if p;

q on condition that p;

q is a necessary condition for p;

etc.

In all these cases, we can express the semantic value of the implication $(\varphi_1 \rightarrow \varphi_2)$ of any two formulas φ_1, φ_2 through the following table:

φ_1	φ_2	$(\varphi_1 \rightarrow \varphi_2)$
0	0	1
0	1	1
1	0	0
1	1	1

In the natural languages such as English, considering a conditional in which the antecedent φ_1 is false is not usual. This can give rise to strange propositions, such as, "If Alexander the Great was a computer engineer, then Aristotle was an astronaut." However, classical mathematics often uses false propositions to set up a situation that can arrive at a proof by a reduction to the absurd.

The Syntax and Models of L0 and L1

As we have seen above, L0 is characterized by atomic (the simplest) propositions. It is also based on a syntax, or structure, of language that is built from an entire alphabet of signs that are governed by the construction rules of propositional logic. The language of first-order predicate logic, or L1, moves beyond the simplicity of atomic propositions. With the language of L1, we can look inside atomic statements for the mathematical functions and relations. In both L0 and L1, the study of syntax, and the building of formal models are among the more technical feats of mathematics. For that reason, I have provided readers who wish to understand this further a set of appendices showing the more advanced mathematics. Appendix 1 discusses the syntax of L0, Appendix 2 presents the construction of the formal models of L0, Appendix 3 analyzes the syntax of L1, Appendix 4 analyzes the semantics of L1, and Appendix 5 presents the numerical systems as the most basic models for L1 (first-order predicate logic).

FINDING THE SEMANTIC "MEANING" IN L0

The remainder of this chapter explores the semantics of L0, or propositional logic, since that can give us insights into the general problem of discovering the meaning conveyed by any language system. When it comes to formal logic and mathematics, we speak of deriving the formal meaning and the real meaning of the propositional language. We make this same kind of double distinction in our natural languages. In the English language, formal meaning is a construction of the human mind. On the other hand, the real meaning is the way the mind connects a particular idea, now associated to words, with the real world of our physical senses.

This activity has always presented a deep philosophical problem: how does the formalization of an idea in the mind correspond to the world outside the mind? It is a question of finding the congruence of logic/mathematics, as constructed in the human mind, with the world of our external experience. Obviously, the formal and the real are profoundly interrelated, but showing just how this takes place is not easy. In the meantime, we have at least distinguished the formal from the real language when we talk about mathematical semantics. The distinction refers to one language that communicates structures produced in the human mind, and the other to transmit experience.

A formal language such as L0 (and L1 as well) is learned through academic training. Similarly, we learn our natural languages by training and exposure to our families and culture in the home as children. Later in life, however, we may learn a second natural language. In some cases, children learn two at the same time because the parents come from two different countries or cultures.

The point here is that while we have our distinct natural languages, and these languages are different in many ways, they nevertheless share certain kinds of common syntax and semantics. These common elements are what mathematicians have tried to incorporate into the formal language of mathematics. This mimicking of natural language is what makes L0 so useful. Its syntax and semantics try to match how our mind also works with natural language. It is the closest thing we have to a universal formal language. A professor of logic can explain L0 and L1 by pointing to the common structures in all natural languages. In turn, natural language can be a metalanguage that explains formalizations such as L0 and L1.

How do we turn a natural language, then, into a formal mathematical one? When we formalize from natural languages, we try to abstract just the common meanings that they share, as happened in the formulation of L0. To take an example, the proposition, "Mary is young *but* Peter is old," does not have the same meaning as "Mary is young *and* Peter is old." The use of "but" in the first puts two facts into opposition, while the use of "and" is neutral. However, in the formalized mathematical language of L0, both conjunctions mean exactly the same thing.

In the formal language of L0, "Mary is young" becomes the sign p, and "Peter is old" becomes the sign q. Adding the conjunctions, both statements are formalized as $(p \wedge q)$. This formalization $(p \wedge q)$ happens because in formal semantics L0 recognizes only the declared truth of both facts. In L0, the conjunction is expressed by a single sign, which is "$\wedge$."

This kind of abstraction allows us to treat language and its meaning as objects of the mind. In the language of L0, for example, we are concerned only with the truth or the falseness of the basic propositions. This abstraction makes it possible to treat propositions in L0 as signs belonging to a signature $\Sigma 0$. We can also express its truth or falseness through a function (or correspondence) that we will call V (for valuation), which makes the value 1 correspond to each sign p of $\Sigma 0$, in the event that p is considered to be true, and the value 0, in the event that it is considered to be false. Both the signature $\Sigma 0$ and the function V are formal objects of the mind.

In addition, because the semantics of L0 that we study here are

classical, we say that every proposition is true or false. As a consequence, we make either 0 or 1 correspond to every formula φ of L0
(see Figure 7.1).

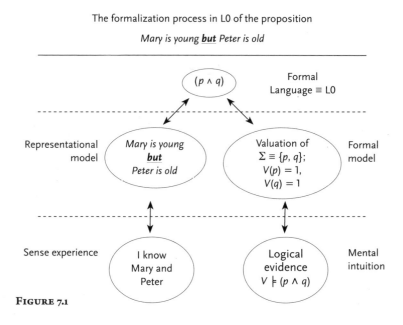

The formalization process in L0 of the proposition

*Mary is young **but** Peter is old*

FIGURE 7.1

For this sort of formal language to stay in contact with the real
world of our physical senses, and to be applicable as a form of communication, the formal language must retain the quality of our
natural language. For human beings, our natural language comes
before we can grasp formal language. The mind is geared first to
our perceptions of the real world and how we articulate that in
words. When I teach mathematics, I need to use both a formal language and a natural language. If I explain mathematics in Madrid, I
explain it in Spanish. In Madrid, in order to be understood, I need
to explain the formulas with examples and comparisons in Spanish. If I present a paper at an international congress, however, I can
do so in English.

When we go from perceiving the real world to talking about it in mathematical language, we must go through a kind of translation, from a natural to a formal language. The models representing reality are first explained in natural language, and thus are not yet strictly formal. For example, we first talk about models of physics, biology, and neuroscience in our natural language, and then we try to put at least part of the language into the formal signs of mathematics.

In this translation from the real-world model to a formal mathematical model, we can think of ourselves as ascending then descending in a total of five stages. The process moves us from empirical reality to the realm of formal logic, and then back to an intuitive conclusion about resolving a truth. We can look at this process by returning to the topic of Mary and Peter. The first step in the process is to observe two empirical facts—two persons called Mary and Peter. The process ends with the construction of a formal model based on the mental intuition of the conjunction between these two individuals. Here are the five steps of our thinking:

1. At the first level we experience real facts. We know Mary and Peter, and we have an empirical perception of the youthfulness of the one and the old age of the other.

2. At the second level, we use natural language to describe reality. By observing the youthfulness of Mary, I make the following affirmation in a natural language: Mary is young. We next observe that Peter is much older. Then we express our observation by saying, "Mary is young, but Peter is old." The proposition that Mary is young but Peter is old is a representational model of observed reality.

3. At the third level, we formalize this perception. This level is academic. In the logic class I teach at the university in Madrid, I would explain to my students that "Mary is young" and "Peter is old" are two atomic propositions. Next, I formalize these through the atomic signs p and q of a signature $\Sigma = \{p,q\}$. I formalize this by

writing that Mary is young ≡ p, Peter is old ≡ q, Mary is young but Peter is old ≡ $(p \wedge q)$.

4. From this level of formalization, we now descend back to the level of the models. But in this case we do not descend to the representational models but to the formal models. The formal models are constructed by our minds. In the case of L0, a formal model is a valuation function V which applies the set $\Sigma = \{p,q\}$ in the set $\{0,1\}$. If we consider, for example, that p and q are true propositions, we define the valuation V by making $V(p) = 1$, $V(q) = 1$.

5. Finally, from the level of formal models, we descend further to the level of logical evidence. The logical evidence is a special type of mental experience. The logical evidence permits us to define the semantic value of the logical connectives. For example, in L0 we have defined the logical symbol ∧ through the following table:

φ_1	φ_2	$(\varphi_1 \wedge \varphi_2)$
0	0	0
0	1	0
1	0	0
1	1	1

This table indicates that the connective ∧ constructs a new true formula from the formulas φ_1 and φ_2 if and only if φ_1 and φ_2 are both true. This logical evidence is not arbitrary and is based, as we have seen, on the very structure of the natural language and on the assumption that every proposition φ is either true or false. According to this table, the valuation V that makes p and q both true will satisfy the formula $(p \wedge q)$; we write this as: $V \vDash (p \wedge q)$.

To observe this process, we have used the analogy of ascending through languages and descending back down into formal models of reality. The goal is to translate empirical facts into formal propositional logic and then apply this to solving a particular problem.

THREE TYPES OF PERCEPTION, LANGUAGE, AND SEMANTICS

It may be evident by now that a formal language that has only syntax (structure) but that lacks semantics (meaning) is cognitively sterile. Nevertheless, in mathematics we often begin by isolating the syntax to clarify its alphabet and its construction rules. In fact the priority given to studying the syntax first shows the privileged status to the formal character of this kind of language.

Once we have the syntax, however, we try to decipher the semantics, or meaning, which is the true source of knowledge. The knowledge we obtain from any language is a combination of its logical signs and the human perceptions that they represent. This unavoidable mingling of languages and perceptions is fundamental, and it takes place even before we try to find semantic rules. The general passivity of perception (we take in data) intertwines with the activism of language (we give out). As the give and take unfolds, experience and language overlap. As we perceive, we formulate language, and as we formulate language, it shapes our perceptions.

All of this is simply to say that semantics must investigate this dynamic process, looking for rules, but also allowing for a good deal of flexibility and interpretation. To handle this complexity, it is helpful to name the three kinds of perceptions that correspond to three kinds of semantics and three kinds of languages. These are very familiar to us by now:

1. *Formal intuitions of a logical or mathematical type.* The formal intuitions of a logical or mathematical type correspond to the languages of logic or mathematics. The formal languages use formal signs that have a formal meaning.

2. *Empirical observations.* The empirical observations correspond to the languages of the empirical sciences. The languages of

the empirical sciences use representational signs whose meanings are real facts and specific data.

3. *Metaphysical intuitions.* The metaphysical intuitions correspond to the languages of metaphysics. The languages of metaphysics use metaphysical symbols whose meaning is real and, therefore, is totally linked to the real facts but does not end in the verification of these real facts. The metaphysical meaning is real and also surpasses and exceeds the empirical meaning.

Based on these three kinds of perceptions and languages, we have three kinds of semantics, which is obvious enough. Our formal semantics corresponds to formal intuitions of a logical or mathematical type. In turn, empirical semantics corresponds to empirical observations. Finally, metaphysical semantics corresponds to metaphysical intuitions.

Most people in our world do not rely on the syntax and semantics of formal language, even though we can say that every normal human mind tends to operate on the principles of logical consistency. The most evident use of formal language in our world today is in academia, where scholars of various disciplines—especially among the sciences—seek a common basis for their research and discussion.

These academic languages are a middle path, a mixture of natural languages and formal elements. These are the languages in which science is usually created, explained, and developed. As every student knows, it is the language of textbooks. And in the modern dialogue between our different human perceptions, and especially between science and religion, this academic language plays an increasingly important role. In the next chapter, however, we begin with a story of ordinary people—six friends—who by discussing the kinds of languages we have reviewed so far show us again their important distinctions, overlaps, and specific uses.

CHAPTER 8
Language and Meaning

THE FORMALIZATION of mathematical language, especially in propositional logic (L0) and first-order predicate logic (L1), has changed our culture. These languages have become the basis of technological and scientific knowledge, and thus social change. Formalized language has brought new kinds of meanings into our world. But how does the public, which is made up of many viewpoints and professions, receive the thinking behind formalized language? How would L1, for example, be approached by a mixed company of modern-day citizens?

To explore that, I offer a story of six friends, five of whom specialize in different kinds of languages. One is an expert in logic. Another knows mathematical language. A third friend is a physicist who does empirical science. The fourth is a theologian, and the fifth friend is a professor of philosophy and metaphysics. The sixth friend has organized their meeting, at which they enjoy discussing the meanings of these five kinds of languages.

The convener's name is Esther. The logician is Jack, and the mathematician is named Maria. Siraj is the physicist. Theresa is the theologian, and Anna the philosopher. Esther has invited them to her home, which has a large picture window looking out upon a marvelous landscape. The landscape includes houses on the outer edge of the city, a valley beyond with a river, a plain of cultivated fields, and beyond that a low mountain range. At the side of the picture window stands a large blackboard. As the convener of the meeting, Esther has made these arrangements, and once the five visitors are seated facing the window and blackboard, she begins the discussion.

First, Esther writes the following sentence on the blackboard:

> The Spirit sleeps in the river,
> dreams in the flowers
> and feels in the birds,
> but only in men and women does the Spirit know that
>   it feels.

Then Esther asks them to explain the meaning of this sentence from their own points of view.

The first to speak is Jack the logician. At the blackboard, he proposes to formalize Esther's sentence to discover its meaning. Naturally, Jack uses the language of first-order predicate logic (L1). He shows how the word "river" can be formalized through a predicate sign with an argument, which could be the letter R. This formalization makes it possible to represent the proposition "x is a river" by writing $R(x)$. Jack writes the following on the blackboard:

$R(x)$: ↔ "x is a river"

Continuing, Jack next explains that the predicate sign with two arguments $SL(x,y)$ can formalize the proposition "x sleeps in y." So he writes the following on the blackboard:

$SL(x,y)$: ↔ "x sleeps in y"

Then Jack formalizes the rest of the basic propositions that appear in the sentence, writing them one by one on the blackboard:

$F(x)$: ↔ "x is a flower"
$DR(x,y)$: ↔ "x dreams in y"
$B(x)$: ↔ "x is a bird"
$FE(x,y)$: ↔ "x feels in y"
$KF(x,y)$: ↔ "x knows that it feels in y"

M(x):↔ "x is a man"

W(x): ↔ "x is a woman"

Finally, Jack formalizes the word Spirit with the sign s and writes:

s: ↔ "Spirit"

After this analysis, predicate by predicate, Jack writes the formula:

$$\forall x \left((R(x) \to SL(s,x)) \land (F(x) \to DR(s,x)) \right.$$
$$\left. \land\ (B(x) \to FE(s,x)) \land (KF(s,x) \to (M(x) \lor W(x))) \right)$$

which he also explains on the blackboard with the following table:

The chain of signs	Means
$\forall x$	All the elements x in the domain of discourse comply in that
$(R(x) \to SL(s,x))$	if x is a river, then the Spirit s sleeps in x
$\land\ (F(x) \to DR(s,x))$	and if x is a flower, then the Spirit s dreams in x
$\land\ (B(x) \to FE(s,x))$	and if x is a bird, then the Spirit s feels in x
$\land\ (KF(s,x) \to (M(x) \lor W(x)))$	and if the Spirit s knows that it feels in x, then x is a man or a woman

Now that Jack has formalized the sentence that Esther presented, the four other friends enter the conversation, beginning with Maria. As a mathematician, Maria says that she could find an infinite number of different meanings that would make Jack's formula true. For example, she could interpret Jack's formula as follows:

R(x): ↔ "x is greater than 4"

SL(x,y): ↔ "x is less than y"

F(x): ↔ "x is greater than 5"

DR(x,y): ↔ "x is less than y – 1"

B(x): ↔ "x is greater than 6"

FE(x,y): ↔ "x is less than y – 2"

KF(x,y): ↔ "x is the square root of y"

M(x): ↔ "x = 2"

W(x): ↔ "x = –2"

s: ↔ 4

Using this interpretation, Maria writes the meaning of the formula on the blackboard:

$$\forall x\,((R(x) \rightarrow SL(s,x)) \wedge (F(x) \rightarrow DR(s,x)) \wedge (B(x) \rightarrow FE(s,x)) \wedge (KF(s,x) \rightarrow (M(x) \vee W(x))))$$

In natural language, Maria's formula can be stated as: "For every integer number x, if x is greater than 4, then 4 is less than x; if x is greater than 5, then 4 is less than x – 1; if x is greater than 6, then 4 is less than x – 2, and, finally, x is the square root of 4 only if x = 2 or x = –2."

She notes that this meaning is true in the domain Z of the integer numbers, but she could find an infinity of different meanings in different formal models. (For interested readers, Appendix 8 explains the formalization of Jack's formula, and Appendix 9 explains the different formal meanings that Maria gives to Jack's formula.)

At this point, Siraj the physicist rises from his chair to approach the blackboard because he has a different view from the one that Maria just presented. For Siraj, Jack's formalization may be correct, but he says that it really explains nothing. The infinite meanings to which Maria referred show the futility of Jack's formalization. Similarly, Siraj says, Maria's many possible interpretations are formal and abstract and detached from the real world. In order to be able to understand the meaning of Esther's sentence, Siraj says they need to determine the real, observable facts of the sentence, which means using an empirical method to confirm whether any of the facts in the sentences are true or false. By now, Jack is pointing out the window to the "real" river, flowers, birds, men, and women in the landscape.

Theresa the theologian has been waiting for her turn in the discussion, and now she offers her point of view. She argues that the meaning of Esther's sentence is found neither in a formal analysis nor in empirical verification. It is not always possible, Theresa says, to empirically verify the meaning of the answers to some questions; many questions in life are about the rich and intangible qualities of things, including the transience of physical phenomena (such as sleep, rivers, dreams, flowers, feelings, birds, men, and women). The interpretation of these qualities can range, for example, between agnostic, nontheistic, and religious viewpoints.

To clarify her position, Theresa distinguishes three levels of meanings: the formal meaning, the empirical meaning, and the kind of meaning that answers metaphysical questions. This last is the symbolic meaning, which can be different for various communities and cultures. The interpretation of a word such as "Spirit" shows the difference, Theresa explains. For some it means a purely human spirit. Others would see it as a nonhuman force. For a Christian theologian such as Theresa, the word could mean the Spirit revealed in Jesus. In the Christian view, moreover, such a Spirit is present in human beings in a different way than it is present in birds. Indeed, for Theresa, the word "Spirit" can have a revelatory meaning, something not at issue with words such as "bird" and "river."

Looking at Jack's formalism, Theresa also argues that there is a distinction between the signs FE(x,y) and KF(x,y) because they represent two empirical realities: one is "feeling" itself and the other "knowing that one feels." To know what one feels can describe a property of a human being only, whereas to feel could be a property of many living beings. Moreover, FE(s,x) and KF(s,x) both contain the sign s. For Theresa, the sign s could have a religious symbolic meaning that, in that case, provides the context of a religious tradition and community that interprets the words in the sentences.

The philosopher Anna expresses her viewpoint last of all. Anna has known Maria, Siraj, and Theresa for some time. And, of course, she often discusses the topics with them in the context of their scientific age. They have discussed, for example, whether there are thoughts of the human mind that a computer cannot reproduce. Anna, along with the others, knows the findings of Gödel (incompleteness) and the Turing machine (undecidability). As a philosopher, she believes it is wise to bypass the computer question and focus on the plurality of systems.

As she takes her turn at the blackboard, Anna argues that it is more fruitful to acknowledge that they are all free to choose formal, physical, and metaphysical viewpoints. These viewpoints do not have to contradict each other, and they could shed light on the others. Then she turns to Maria (the mathematician) and, gesturing out the window to the city, asks whether anyone living in the houses would interpret the meaning of Jack's formalized logic as she did (through a model of the integer numbers)?

Naturally, Maria responds by saying that only a mathematician would give an interpretation like hers. At the same time, she admits that in this mathematical approach, the possible interpretations are almost unlimited, allowing mathematicians to choose even the most bizarre system. In mathematics, it seems, all was possible. At this point, Anna asks Siraj the physicist about this open-ended viewpoint. Siraj says it is meaningless to him, for he is most confident in what can be determined by empirical measurement. In fact, Siraj accepts the possibility that the number of particles in the universe is finite, and therefore, the number of interpretations of the world must also be limited.

As an empiricist, Siraj also points out that we have evidence that the way human beings know is different from how birds know; human beings produce the scientific practice of physics, while birds cannot do this. In fact, a human being not only feels but knows that he or she feels and can turn this into an objective perception. But for all this, Siraj also acknowledges that, as an empiricist, he simply accepts this as a given fact of the universe. He does not need to ask why human beings have a power to know differently than how birds know. This is a mystery. If Theresa approaches this mystery about the Spirit with symbolic language, Siraj remains agnostic. He prefers to stick with empirical answers alone and avoid other speculations.

In this conversation, Anna has the final word. She says that while much human experience is hard to express, we nevertheless try to capture it by use of a language. We formulate these languages differently in logic, mathematics, science, theology, and metaphysics. Mathematics and natural science are especially concerned about the precision of their languages. In metaphysics and religion, Anna says, the meaning of the language is more important than its technical precision, because what religion or metaphysics can explain cannot be easily expressed, and thus it turns to symbols of meaning.

As the meeting at Esther's home concludes, the participants agree that they have the most common ground on formal logic. That is the basis for a consistency of their thinking and discussion. This formal logic, moreover, does not necessarily contradict the empirical or metaphysical approaches to perception and language. It is equally clear, however, that formal language is not sufficient in itself to satisfy the empirical interests of Siraj, the physicist. Nor is the empirical language sufficient to satisfy the symbolic and cultural interests of Theresa, the theologian. She needs a language of great richness, which includes references to the transcendent.

Formal, Empirical, and Metaphysical Meaning

Our six friends who grappled with the meaning conveyed by formal, empirical, and metaphysical languages are not the only ones preoccupied with these sorts of questions. In the remainder of this chapter, we consider how our efforts to organize language as a carrier of meaning is central to human progress, whether in scientific advancement or the perpetuation of our value systems.

Formal Meaning

While mathematics is our most formal language, it too was created to convey meanings about the world. As time passed, mathematicians inevitably mixed their formalism with natural language, trying to retain some of the natural meanings of those languages. Today, however, we have tried to completely formalize mathematical language—and yet keep its meaning intact. We can briefly review how this process begins.

In order to formalize a proposition of mathematical language in first-order predicate logic $(L1)$, we must construct a formal interpretation (S,σ) that is adapted to the proposition we wish to formalize. To do this, we must compose (S,σ), which is made up of a structure (S) and a valuation (σ) of the variables. The structure S is composed of, first, a domain of the discourse D, and next a formal meaning in D $(c^D \ldots ; f^D \ldots ; P^D \ldots)$ of the signs of a signature $\Sigma1 = \{c \ldots, f \ldots, P \ldots\}$. In summary, the valuation of the variables σ is an application of an element $\sigma(x)$ of the domain of discourse D to each variable x.

The above process of constructing (A,σ) is a skill of the human mind, which has developed over the centuries, making (A,σ) a formal model that interprets mathematical intuitions. Current mathematics has innumerable possible interpretations (A,σ) formally constructed. In order to formalize a usual proposition of mathematics, it is not normally necessary to construct a new formal interpretation; we can take it from among those existing in current mathematics.

We interpret the meaning of the words of the proposition we wish to formalize using logical signs and signs of the signature. We construct a formal proposition whose meaning interprets in (A,σ)

the meaning of the proposition that we wish to formalize. To take this further for interested readers, I have presented an example of a mathematical formalization in Appendix 10.

As we have seen, the construction of the formal meaning of a proposition is a process involving the translation of a nonformal language to a formal language. The semantics of the nonformal propositions is not totally defined. It may be imprecise and may give rise to several interpretations. In the formal languages, the meaning is given by a function that is formally defined in accordance with the rules of language.

Empirical Meaning

In empirical science, the formal signs of mathematics stand for observed facts in the physical world. Hence, the meaning is not simply formal, but it is a meaning about physical objects. The empirical sciences have also branched into many different fields; therefore, while they may use the same mathematical notations, these notations refer to different material objects of each science. The signs have a different meaning when they are applied in biology to designate living organisms, in physics to material objects, or in neuroscience to the human nervous system.

We can return to Galileo Galilei, the seventeenth-century Italian scientist, to see how formal signs (as in mathematics) were applied to his empirical models of moving objects. In his experiments, Galileo observed that the distances traveled by a falling body were proportional to the square of the time elapsed.

He did not know the universal law of gravity. But he could make a model capable of representing his observations regarding falling bodies. In Galileo's model the following equations were verified:

$$d_1 = k \times 1^2$$

$$d_2 = k \times 2^2$$

$$d_3 = k \times 3^2$$

In this model, d_1, d_2, and d_3 are signs that represent distances traveled and 1^2, 2^2, and 3^2 are signs that represent times elapsed. Based on this model and consistent with Galileo's observations as published in physics textbooks, we find the classical law of mechanics $d = \frac{1}{2} g \times t^2$ that regulates falling bodies attracted by the gravitation of the earth. This law has a clear, real meaning in classical mechanics, according to which the constant g represents the particular value of the gravitational constant on the earth's surface. Galileo's empirical observations have contributed to the creation of a physical model of reality that we call classical mechanics. In turn, the model of classical mechanics is explained in the academic language of physics, which includes mathematical formulas such as $d = \frac{1}{2} g \times t^2$.

Metaphysical Meaning

How do we create a formal language to convey metaphysical meaning? Metaphysics asks about the ultimate principles. This is an age-old and inexhaustible type of question. The remains in prehistoric tombs testify to metaphysical questions and answers concerning the permanence of life.

In every cultural period thereafter, the responses to the basic metaphysical questions have acquired different nuances. These nuances were often based on different scientific environments. When mathematics became a quest for certainty, for example, so did religious systems. That is why the Pythagoreans in ancient Greece extended the absoluteness they found in numbers to metaphysical absolutes that they believed governed the universe. The Greeks were not able to reflect mathematically on mathematics, as we do with metamathematics today. And today our metaphysics reflects the nuances of metamathematics, with its emphasis on open systems.

If we can link mathematics and metaphysics, we should also be able to link metaphysics with science, which relies on mathematical language. There is something of the absolute in empirical science as well. We can't really ask metaphysical questions apart from empirical reality. If anyone doubts that empirical science has tried to make absolute claims, we can recall our earlier account of David Hume, who was so confident in science that he urged that all books on divinity and metaphysics be burned.

We might agree with Hume that metaphysics and religion in each age ignore the empirical sciences at their peril. However, science and mathematics are proving—now more than ever—their own openness and pluralism, allowing for many kinds of overlap with the concerns of metaphysical viewpoints. We are wide open today, for example, on the ultimate principles of knowledge. How do we obtain final knowledge, and what kinds of knowledge are the best and most useful?

This question returns us to the way the six friends tried to determine true and useful knowledge when confronted with the same sentence of information:

> The Spirit sleeps in the river,
> dreams in the flowers
> and feels in the birds,
> but the Spirit only knows that it feels in men and women.

As the logician, Jack was the first of the group to try to give meaning to this sentence. He calls this the formula φ, and presents it on the blackboard as follows:

$$\varphi : \forall x((R(x) \to SL(s,x)) \land (F(x) \to DR(s,x))$$
$$\land (B(x) \to FE(s,x)) \land (KF(s,x) \to (M(x) \lor W(x))))$$

The essence of Jack's formal logic is that the sentence is extremely open to a wide range of meanings (for more on this, see Appendix 10). Even our most reliable, valid, and simple language thus proves to be open, which allows Maria, the mathematician, to use it however she likes. The openness of Jack's system enables Maria to formulate her own meaning by specifying a domain of discourse, A, and a set of whole numbers Z based on her own arbitrary choosing. (For more on this, see Appendix 8.)

Maria can choose infinite different structures in which to interpret Jack's formula. An infinite number of different sets can be a domain of discourse in which Jack's formula is true. Maria can make φ true in an infinite number of ways, and from among those countless meanings, she simply can choose one that she prefers.

Perhaps as a result, Siraj the physicist does not find Maria's formal meanings interesting, even though they are valid. He can only find meaning in relating the signs to empirical experience. For Siraj, these systems of formal signs can only have meaning if they are stated as hypotheses that physical experiments can test, trying to falsify the hypothesis. In the case of formal logic, Siraj does not see any way to disprove it, and so it is outside of empirical science.

Among the six friends, Theresa the theologian and Anna the philosopher come to the discussion from an entirely different viewpoint. While they appreciate logic, math, and science, they believe that the more important kind of meanings in human language come not through signs but through metaphysical symbols. For Theresa and Anna, the meaning of the word "Spirit" is not a matter of logical existence or a physical object. It is a metaphysical meaning about something unique in the human being not found in nature, or, given Theresa's background, about a reality spoken of in the Christian tradition.

That the six friends are indeed friends is a good thing, for they otherwise are giving a different priority to different languages, often a source of conflict in the world. The wisdom to be found here is in the acceptance of pluralism of languages. To some extent, there is also a common root. For example, in metaphysical propositions, there is also a formal meaning (logic) and an empirical one (based on experience in the real world, such as history, tradition, and community).

We began our story with Jack because even metaphysical questions must retain a formal and empirical language. The signs and symbols of language complement each other. One can build upon another. Contradiction is not inevitable or necessary. Surely, Theresa and Anna know this if they want their religious or metaphysical projects to stay in touch with reality.

We can end this story of the six friends by recalling the distinction made in the first chapter of this book (see Figure 1.2), which looked at the perceptions, models, and languages of mathematics and empirical science. Now we can apply these three levels to metaphysical meaning as well, as suggested by Figure 8.1.

We have seen in the history of mathematics and empirical science that there always remains a side of reality that is unknowable and inscrutable—a mysterious background. Our desire to apprehend that mystery is what drives our intellectual curiosity in every culture and age, and this desire often is the force behind mathematics and science as well.

But such curiosity is unique to the human domain, beyond what machines or formula can penetrate. Only men and women perceive this mysterious background. Computers can prove the incompleteness of arithmetic, but they do not feel the thrill of this quest for knowledge and proofs. The thrill raises the great metaphysical questions in our minds. As we see in the next and final chapter, we try to answer metaphysical questions with metaphysical and religious models, communicating about them in symbolic languages.

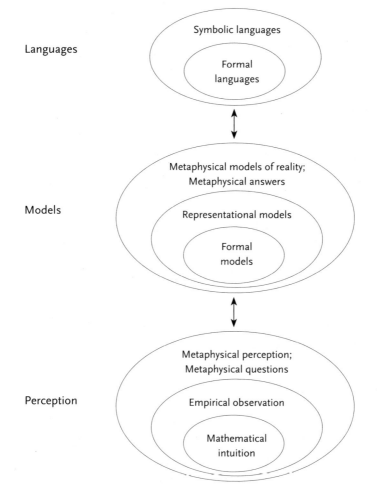

Languages

Models

Perception

FIGURE 8.1

CHAPTER 9
Science, Language, and Religion

SCIENCE AND ITS LANGUAGE have had a profound influence on our epoch. Logic continues to be at the heart of our natural languages, and mathematics retains its privileged status as a public language. One simple example of this is how, around the world, we successfully translate a wide array of complex mathematics textbooks into Chinese, Hindi, Malagasy, and other dialects regardless of the language of the original author of the book.

In all of this, our hope has been that logic, mathematics, and science can lift us above our subjective biases to a plateau of objective knowledge and perspective. We seek a public language around the globe. Based on the same rudimentary logic of mathematics, we have developed technology. Thanks to technology, we have cars, aircraft, computers, and refrigerators. We have instruments for scientific observation and medical treatments. Technology has given our present-day culture progressively more power over nature.

But perhaps in this area of technology the wider public is now questioning the powers of our formal and scientific languages. These languages allow us to do many things, but they have also made some aspects of our lives machinelike, automated, and routine. The computer, of course, is the great symbol of where the world may be headed. The computer is a key part of modern industry, and industry is driving such topics of contemporary distress and debate as climate change, the loss of natural resources, and ecological imbalances. Plato once worried about the modernizing influence of reading (rather than discussion) on young people, and we are seeing a similar concern today in a growing criticism of the language (and power) of science.

Scientific Language under Fire

Fortunately, human beings have the ability to stand outside of their natural languages and look at them critically. This is what we have achieved in mathematics, for example, with the rise of metamathematics: we subject math systems to the very principles they are made of to look for error as well as consistency. It is remarkable to see how a trained monkey, for example, can carry out methodical activities, relating one number to another number and expressing that with sounds. But neither a monkey nor a computer—yet, at least—can self-reflect on this type of activity, its meaning, or the language used.

Throughout the twentieth century, scientists and philosophers looked at the historical, psychological, and sociological roots of scientific language and method. For example, many kinds of scientific language clearly have developed in history. Each one was based on a different way of knowing—a different epistemology. The language of quantum physics is not reducible to the language of large-scale physics. Quantum physics can only forecast the probable conduct of quantum particles. With the language of macroscopic mechanics, however, we can forecast with certainty the movements of bodies.

One influential work of our time that has pointed out these different kinds of scientific languages, and how they often rival each other, is *The Structure of Scientific Revolutions*, by the physicist and philosopher Thomas Kuhn (d. 1996). He argued that scientific revolutions take place when anomalies arise in certain conventional systems of science and scientific language, and these anomalies are rebelled against by a new system of ideas and language. This process occurred during the Copernican revolution in astronomy, the Darwinian revolution in biology, and the quantum physics revolution.

Beginning with the language of observation, all of these revolutions have eventually changed the entire scientific culture of the discussion, often with new languages.

After a long history of science, therefore, we now have a plurality of scientific communities, each with its own chief hypotheses and its own special language. Even though formal logic remains basically at the core of this development, the language systems can multiply based on fundamental disagreements about logic or about the problem that a system is trying to resolve. Actually, nothing is wrong with this critical character of science, for that is what allows us to test our hypotheses and move closer to valid and useful findings. The noted philosopher of science Paul Feyerabend (d. 1994), for example, stirred a lively debate about the limits of scientific culture with arguments about the frequent inadequacy of its language, writing such popular books as *Against Method* (1975) and *Science in a Free Society* (1978).

Scientific Pluralism

We are left, then, with a plural scientific culture. It has a commonality of method (empirical) and common tools for analysis (logic and mathematics), but it is also divided by its interests and application of these tools. The sciences are diverted to disparate observations of nature. Some look at astronomical objects, others at biological systems, and still others, in the social sciences, at human behaviors. The result is a growing plurality of scientific languages, often with a substantial lack of communication among the different fields.

The most troubling question for science is the following: how can we trust scientific language to be objective and universal? Some scientists have spoken openly of the limits of objective language when it comes to the subjective experiences that inevitably influence scientific research. The study of mental perception by human beings is a telling example. When we scientifically observe mental perceptions, we find subjective qualities of these perceptions, which we call *qualia*. By introspection, we can perceive color, pain, and other sensations, but these *qualia* also have a strong subjective component. We can objectively state that the perception of a color depends on the object we observe, such as a blue car, a green forest, or a rosy sunset. But how we feel about those objects is a matter of subjective interpretation. This problem also arises in other fields of science that require human perception, such as quantum physics.

As we saw in earlier chapters, this plurality of languages also extends to mathematics, which is a very human science. Some mathematicians are committed to classical mathematics with its logical principle of the *tertio excluso*, according to which every formal proposition is true or false. Other mathematicians, on the other hand, hold only that a formal proposition is true or false when we have effectively demonstrated its truth or falseness. Some mathematicians admit the existence of sets with infinite elements. Others only admit the existence of sets that have an arbitrarily large number of elements.

FINDING UNITY THROUGH A THEORY OF SYSTEMS

Nevertheless, the quest for unity in the sciences has not ended. At this time, the most promising vehicle for that unity continues to be our simplest scientific language: formal mathematical language. This quest is never easy, since even the basic logic of formal mathematics must operate in a system. As seen in the examples in this book, the formal systems of mathematics and the real systems of the empirical sciences remain in tension, the tension between language and experience.

Our most effective way at present to approach science as a whole is by using a theory of systems, which has become a formal area of study. In the theory of systems, we can look for common properties in all the different systems. We would call this a metatheory, since it tries to stand outside a particular system, and it has also tried to provide a metalanguage that can talk about common properties in the disparate systems.

For example, one of these common properties can be called the adaptive capacity of a system. Such systems are capable of adapting themselves to changes that occur in the environment. Living beings are adaptive systems because they react to the changes around them, often adapting very well. If a plant finds light in a part of the room, it adapts and directs its branches to that source. The stock market is an adaptive system because it reacts to political and economic changes. Political parties and human communities are all adaptive systems. Science can thus talk about common characteristics and seek a common application of logic and language.

Another property common to some systems is their capacity to self-generate, called autopoiesis. Living beings, in general, are autopoietic. We have all felt the restorative effects of a good night's sleep. A third property that the theory of systems considers is called entanglement. First observed in quantum physics, particles of a common origin are entangled in such a way that they describe and define each other, even though there finally is no physical connection between them. Entanglement has turned out to be a very effective principle of analysis in other systems, such as human neuroscience and psychology.

Although the theory of systems began in natural science, it is being applied to mathematics as well. The geometry of Euclid is a formal system, for example, so it is subject to a theory of systems. Similarly, first-order predicate logic is a formal set of systems whose alphabet comprises logical signs $(\forall, \exists, \neg, \wedge, \vee, =)$ and variables $(x_1, x_2, x_3, \text{etc.})$. Then, there is a specific set of signs proper to each first-order system. For example, $\Sigma \text{ar} = (0, 1, +, \times, <)$ is a set of signs proper to first-order arithmetic. In general, all mathematical systems are characterized by a formal language (syntax) and by their formal meaning (semantics). So they can be subjected to a theory of systems.

The mathematician Alfred Tarski (d. 1983) pioneered this analysis by proposing a theory of models. Because computers can manipulate the formal languages of logic and mathematics, we can create these formal models with relative ease and precision. The theory of models looks for a common type of semantics among all of these formal systems by using a constructive definition of the truthfulness or falseness of formal propositions in a given system. Meanwhile, researchers are also using alternative semantics to the semantics of the theory of models. They are experimenting with fuzzy semantics (which has several truth values), games semantics, and also what we call the proof theory. All of these look for commonalities in our diverse mathematical and scientific systems.

We began this book by talking about the formal language of mathematics and the representational language of natural science, which are the basis of our mental and empirical systems of research. The conclusion we have arrived at is that both of these kinds of systems, formal and real, are open. In other words, some things about these systems, though having internal consistency, cannot be resolved. The main reason for this lack of resolution is that systems develop properties that do not arise directly from their most basic foundations. As the old phrase goes, a system turns out to be greater than the sum of its parts.

Therefore, any sufficiently complex formal system usually is an open system. One way that mathematicians have realized this is through the modern theorems of incompleteness and undecidability. In natural science, moreover, open systems are increasingly characterized by our study of quantum physics, biology, and neuroscience. Each of these natural systems witnesses a variety of emerging properties that cannot be simply built up from simpler properties in the system. Mathematics and natural science have both led us to ponder metasystems and metalanguages in order to understand properties that emerge, often unexpectedly, from the heart of a system. As noted earlier, one of these properties can be self-generation, or it can be adaptation. Either of these can now be studied in nature or in a computer program.

With the theory of systems, we have found a new ontology—a new theory of existence, that is—that allows us to continue to seek a unity among the scientific disciplines. The ontology of system can be written in a formal language so that it can be understood by a computer as well. However, an open system always defies complete control and resolution, and so it always puts limits on the languages of mathematics and science. In an age of open systems, we are still forced—and perhaps asked more than ever—to ponder metaphysical questions.

THEOLOGY IN A SCIENTIFIC CULTURE

The openness of mathematics and science is good news for the language of metaphysics, religion, and faith. Mathematics and natural science must begin with an assumption, and it is an assumption they choose. That is to say, both disciplines put a certain faith in their assumption and then work outward from that (which we typically call "deduction," and is especially characteristic of logic and mathematics). The process is not too different in metaphysics, which includes religion. The medieval scholastics defined theology as *fides quaerens intellectum*, "faith seeking understanding." Similarly, we could say that modern science is *perceptio metodica quaerens intellectum*, or "perception and method seeking understanding." Faith (*fides*) and perception (*perceptio metodica*) are parallel experimental ways of obtaining knowledge of human experience, nature, and history.

Science takes in reality through methodical observation. Theology takes in reality through faith. In both cases, the human mind seeks to understand through formulation in a language and the logical structuring of the language. Scientific theories and theological theories are both produced with the common instruments of language and logic. Currently humanity finds many answers in science that our ancestors searched for, and resolved, in religion. This has allowed modern faith to free itself from preoccupations with physical science, a topic that is often not relevant for, or even alien to, what the faith experience is all about.

In the meantime, the languages of science and metaphysics can be transcultural. They can speak to all people despite the barriers of our different natural languages. In science, it is the language of 2 + 2 = 4 or the laws of gravity. In metaphysics and religion, it is the language of absolute reality, such as a transcendent Creator or principle. The idea of God may be expressed within the context of a culture, but in principle, that culture cannot limit such ideas to exclude other human beings who intuit the same higher reality. What is more, in most of our religious traditions, we believe that the Creator reveals this intuitive knowledge to all men and women. For the believer especially, God is seen as actively present in the world. The life of the believer can become a response to a sense that life is a gift, not just a deduction.

Science and religion have shaped our cultures. Our current challenge is to keep a discussion going on between these different kinds of perceptions and languages. Indeed, the history of Christianity (and Judaism and Islam, for that matter) can be viewed as a series of responses to scientific cultures over the ages, from the Hellenistic time through the Enlightenment up to the present age of quantum physics. Christianity today grapples with these same challenges on the scale of a global scientific-technical culture. The challenge, stated by many modern thinkers, was well articulated by Pope John Paul II in a 1988 letter to the director of the Vatican Observatory, George V. Coyne, SJ:

The hylomorphism of the natural philosophy of Aristotle, for example, was adopted by medieval theologians in order to use it in the examination of the nature of the sacraments and the hypostatic union. This did not mean that the Church judged the truth or falseness of the Aristotelian conception as this is not its concern. It meant that this was one of the grand conceptions offered by Greek culture, which needed to be understood, taken seriously and its value to illuminate several areas of theology verified. Today theologians could ask whether they have carried out this extraordinarily difficult process as regards science, philosophy and other areas of contemporary human knowledge with the perfection with which the medieval masters did so.

In the past, religions tended to be limited by culture, exposed to only a single culture's internal scientific and philosophical worldview. Now, every religion faces a global scientific-technical culture. This gives an important role to the great religions—such as Judaism, Christianity, Buddhism, and Islam—to understand each other and to adapt themselves as metaphysical options to a scientific age.

A primary way to do this is for the great religions to stay conversant with scientific language, which helps them share scientific culture as well. Religions can do this quite safely by recognizing that faith cannot be deduced from empirical knowledge. To quote Augustine, the transcendent cannot be deduced from the immanent (*"Si comprehendis non est Deus"*). In effect, the silence of scientific language toward the God question helps purify religious faith, allowing the believer to find harmony between the laws of the world and the presence of a Creator.

Of course, voices in history have declared that science and religion, because they have opposing perceptions and language, are in mortal conflict. As this book has tried to show, the warring ramparts are not as firm as either science or religion, in their more dogmatic eras, once had believed. The world and its natural systems are open, and the transcendent is a logical conclusion we can draw from our consistency of thought. Mathematics and science try to answer how things are. Metaphysics and religion try to answer why the world is the way it is. We are wise to realize that these questions complement each other, which encourages the delight and curiosity we feel about life. I cannot ask why I am in the world if I am not interested in knowing how I am in the world.

A central argument of this book is that the language of mathematics holds a privileged status in human affairs. It is a kind of public language that allows us, as best we can, to try to achieve objectivity and certainty. It is more than that as well. Mathematics manifests its knowledge between the extremes of the absolute and nothing. It helps us navigate between the tendency to be subjective or nihilistic, and our tendency to be overconfident and dogmatic. Mathematics shows us that there are certainties—including a kind of logic that makes our natural languages possible—but there also is incompleteness and openness.

For absolute knowledge we must turn to metaphysics and its particular language of symbols in the context of tradition and community. The very fact that this search continues illustrates a kind of sublime, disinterested, and universal consistency in the human mind. This was what prompted some philosophers of the past (in the ontological argument) to try to prove God's very existence by logic. We do not need to go that far today; it is enough to acknowledge the consistency of our logical search.

Reconciling the "Magisteria" of Science and Religion

As a final reflection, I would like to draw on some of the resources of my own tradition of Catholic thought. In that long tradition, we have spoken of the deposit of faith and truth, conveyed by the church in councils, documents, and wisdom, as "the magisterium." It is popularly called the teaching authority of the church. Not long ago, this topic gained wide attention through the writings of the noted American paleontologist Stephen Jay Gould.

In the 1980s, Gould had been invited to Rome for a conference on science organized by the Pontifical Academy of Sciences. He was apparently taken by the idea of a magisterium, or teaching authority. This experience in Rome, he later said, prompted him to propose a model of the relationship between science and religion called the Non-Overlapping Magisteria, or NOMA. As he wrote,

> The lack of conflict between science and religion arises from a lack of overlap between their respective domains of professional expertise—science in the empirical constitution of the universe, and religion in the search for proper ethical values and the spiritual meaning of our lives. The attainment of wisdom in a full life requires extensive attention to both domains—for a great book tells us that the truth can make us free and that we will live in optimal harmony with our fellows when we learn to do justly, love mercy, and walk humbly. . . . The net of science covers the empirical universe: what it is made of (fact) and why it works this way (theory). The net of religion extends over questions of moral meaning and value. These two magisteria do not overlap, nor do they encompass all inquiry (consider, for starters, the magisterium of art and the meaning of beauty).

In his desire to avoid conflicts between science and religion, Gould allowed religion its own kind of authority, ethical and moral. But in his desire to give the scientific magisterium a higher ranking in knowledge of the "real" world, Gould was puzzled by the continued effort in Catholic thought to say that science and religion do indeed have truth in common. He was particularly puzzled by the 1996 statement of Pope John Paul II titled, "Truth Cannot Contradict Truth."

As an answer to Gould's puzzlement, I would like to offer an alternative to NOMA. I would like to propose the model of a relationship between science and religion called the Non-Symmetrical Magisteria, or NOSYMA. Here, science and religion cannot be separated. Their relationship is complementary, but it is not a symmetrical relationship. Instead, we can say that religious knowledge needs science, while science may do without religion. In effect, this asymmetry is a plus for science by making it autonomous, but it is also a plus for religion by endowing religion with a more comprehensive vision.

In this way, I would argue, the two magisteria cannot be separated. Furthermore, they both use a common language that is characterized by the desire to be logical and consistent. Logic and consistency are what make mathematics the language of science par excellence. The language of theology has also tried to be logically consistent with the language of science.

However, faith cannot close its eyes to mathematics and the empirical sciences. I can separate mathematics from theology, but I cannot separate theology from mathematics. Mathematics and the empirical sciences are independent of religious beliefs, but theological reflection cannot do without mathematics and the empirical sciences.

This asymmetrical model corresponds to how human beings seem to live their lives and how we develop our scientific and religious knowledge. My belief that science and religion complement each other in an asymmetrical way also arises from my own intellectual and spiritual journey in life. I know very good scientists and mathematicians who are believers and others who are nonbelievers. Since I have chosen faith, I can certainly separate my mathematical work from theology, but I cannot separate theology from what I learn in mathematics.

In my personal history, God has become present for me in the world through his Son Jesus. But I can only know Jesus if I know the world. The knowledge of the world for me has been an essential part of my religion. Therefore, my religion needs science in order to continue to be human and to unite people with God.

We also need motivation, hope, and a vision in life, which is something religion offers that science, in proper humility, does not claim to provide. Religion has motivated my scientific search, for example. But I am also aware that science can reject my religion. I have known scientists who have been helped in their work by religion, drawing motivation and inspiration from its wellspring. I've also known others who have renounced religious inspiration in the name of science.

Either way, I have decided to view my faith as a gift. As a member of the Society of Jesus, I view Jesus as a companion in my voyage between the absolute and nothing. This is contemplative relationship, not a research finding. For me, faith is a gift received. And I hope this little book may be taken in the same spirit as I have tried to show how our languages unite us more than they divide us.

APPENDIX 1
Syntax of Propositional Logic

PROPOSITIONAL LOGIC studies the consistency of language at the level of propositions. A proposition is a statement to which we can attribute a value of truth or falsehood. In this appendix we present the formal language of propositional logic (syntax). In Appendix 2 we present the formal meaning (semantics) of this language.

The syntax of propositional logic provides us with an alphabet of basic signs and a set of construction rules. Through the construction rules we produce all the formulas of the language of propositional logic from the basic signs of the alphabet.

The Signs of the Alphabet

A signature $\Sigma 0$ for the propositional logic is any decidable set of signs. A set is decidable if we have an effective criterion to recognize whether a given sign belongs or does not belong to it. All the finite sets are decidable. An infinite set can be decidable or not decidable. We will designate the elements of a decidable signature $\Sigma 0$ for the propositional logic with the letters p, q, r, . . .

An alphabet A0 with the signature $\Sigma 0$ is the union of the following sets of signs:

— A signature $\Sigma 0$.
— The set of logical signs $\{\top, \bot, \neg, \wedge, \vee, \rightarrow\}$
— The set of auxiliary signs $\Sigma 0 U\{\top, \bot, \neg, \wedge, \vee, \rightarrow\}U\{(,)\}$

Convention: In the following we adopt the convention that the sign $\square$ represents any of the binary connectives $\wedge, \vee, \rightarrow$.

Formulas of Propositional Logic

Given an alphabet A0, we construct the set L0 of the propositional formulas through the rules (At), (¬), and (□):

(At)   Any sign p of Σ0 is an atomic formula.

(¬)   If φ is a formula ¬φ is also a formula.

(□)   If φ₁ and φ₂ are formulas (φ₁ □ φ₂) is also a formula.

The rule (At) makes it possible to construct the set of all the atomic formulas of L0. The rules (¬) and (□) make it possible to construct all the compound formulas.

Example

Supposing that Σ0 = {p,q,r}, through the rules (At), (¬), and (□) we can construct, for example, the proposition ((p ∧ q) → (¬p ∨ q)) in six steps:

Step	Constructed	Rule	Formulas used
1	p	(At)	
2	q	(At)	
3	(p ∧ q)	(□)	1, 2
4	¬p	(¬)	1
5	(¬p ∨ q)	(□)	4, 2
6	((p ∨ q) → (¬p ∨ q))	(□)	3, 5

Given a decidable signature Σ0 the alphabet A0 is also decidable. Each formula is a finite succession of elements of the alphabet A0. However, not all the finite succession of signs of A0 are formulas of the language L0. For example we have verified that if Σ0 = {p,q,r},

SYNTAX OF PROPOSITIONAL LOGIC : 135

then $((p \wedge q) \rightarrow (\neg p \vee q))$ is a formula of L0; we can also verify that $(\neg \rightarrow p)$ is not a formula of L0 as it cannot be constructed through the rules (At), $(\neg)$, and $(\square)$.

The Problem of Decision

A subset of a set is decidable if we have an effective procedure to determine whether a given element belongs to the subset or not. For example, the subset of prime numbers is decidable because we have an effective procedure that can be executed with a program through which we can decide whether a given natural number is a prime number or not.

We can prove that the formulas of L0 are a decidable subset of the set of all the possible finite successions of signs of A0. The proof is carried out by constructing an algorithm through which we can decide whether a given succession of signs of A0 is a formula of L0 or not.

APPENDIX 2
Semantics of Propositional Logic

THE LANGUAGE L0 consists entirely of formulas that represent statements whose meaning is true or false. The semantics of L0 allows us to determine by mechanical means the value of truth or falsity of each statement in a formal model. The following is an example of how this semantic system is constructed:

A valuation of a signature $\Sigma 0$ is an application V of each sign of $\Sigma 0$ to the set $\{0,1\}$, where 1 represents truth and 0 falsehood.

$$V: \Sigma 0 \rightarrow \{0,1\}$$

A valuation of the signs of $\Sigma 0$ defines a formal model of the language L0.

In order to determine whether a given formula of the language L0 is satisfied by a given valuation V of $\Sigma 0$, we construct an application mapping all the formulas of L0 to the set $\{0,1\}$. Thus, the set of all the formulas that are true in a given valuation V are mechanically determined.

For example, given the signature $\Sigma 0 \equiv \{p, q\ r\}$, we can define the following valuation:

Atomic proposition	Semantic value
p	1
q	0
r	1

.

In this case we write $V(p) = 1, V(q) = 0, V(r) = 1$.

The signature $\Sigma 0 \equiv \{p, q, r\}$ has eight different possible valuations, as represented by the following semantic table:

p	q	r
0	0	0
0	0	1
0	1	0
0	1	1
1	0	0
1	0	1
1	1	0
1	1	1

In general, if the number of elements of $\Sigma 0$ is n, the number of different possible valuations (formal models) of $\Sigma 0$ will be 2^n.

Now, for each valuation $V: \Sigma 0 \to \{0,1\}$, we define an application $[\ ]^V$ which assigns to each formula φ of L0 its semantic value $[\varphi]^V$, where $[\varphi]^V = 0$ or $[\varphi]^V = 1$.

$$[\ ]^V: L0 \to \{0, 1\}$$

In order to define the application $[\ ]^V$, we use the auxiliary Boolean functions $v_\neg, v_\wedge, v_\vee, v_\to$ associated respectively to the connectives $\neg, \wedge, \vee, \to$.

The values of $v_\neg(x)$ are given by the following table where, when x takes the values 0,1, $v_\neg(x)$ takes the contrary values 1,0, respectively.

x	$v_\neg(x)$
0	1
1	0

The following table represents the values of $v_\wedge(x,y)$; $v_\vee(x,y)$; $v_\to(x,y)$ corresponding to all the possible values 0,1 of the Boolean variables x,y:

x	y	$v_\wedge(x,y)$	$v_\wedge(x,y)$	$v_\to(x,y)$
0	0	0	0	1
0	1	0	1	1
1	0	0	1	0
1	1	1	1	1

Remembering the agreement that the sign □ represents any of the binary connectives ∧, ∨, →, given a formula φ of L0, and a valuation V: Σ0 → {0,1}, we calculate the semantic value $[\varphi]^V$ through a construction defined as follows:

— Base of the construction:

We define the semantic value of the atomic formulas:

(At): $[\top]^V = 1$; $[\bot]^V = 0$; $[p]^V = V(p)$ for every p of Σ

— Construction steps:

$(\neg)$: $[\neg\varphi]^V = v_\neg([\varphi]^V)$
$(\square)$: $[\varphi_1 \square \varphi_2]^V = v_\square([\varphi_1]^V,[\varphi_2]^V)$

Formal Models of L0, Satisfactibility

If $[\varphi]^V = 1$, we say that V satisfies φ or that V is a model of φ, and we write it thus:

$V \models \varphi$

If $[\varphi]^V = 1$, we say that V does not satisfy φ or that V is not a model of φ.

APPENDIX 3
Syntax of First-Order Logic

THE SYNTAX of first-order predicate logic includes new signs for variables (Var), functions (F), predicates (P), and existential (∃) and universal quantifiers (∀). This syntax is also specified by a decidable alphabet of signs and rules for the construction of formulas.

The Alphabet

The A1 alphabet of first-order predicate logic is an extension of the A0 alphabet of propositional logic and includes the following sets:

— A decidable signature of signs $\Sigma 1 = F \cup P$, which is the union of a set $F = \{f,g,h, \ldots\}$ of function signs and a set $P = \{P,Q,R \ldots\}$ of predicate signs. The function signs with zero arguments are called constants, and we designate these with the letters c,d,e, . . . The predicate signs with zero arguments are proposition signs, which we designate with the letters p,q,r, . . .

— The logic signs are:

The propositional connectives $\{\perp, \top, \neg, \wedge, \vee, \rightarrow\}$
The quantifiers ∀, ∃
The sign of equality =

— A numerable infinite set $Var = \{v_0, v_1, \ldots, v_i, \ldots\}$ of variables. We agree that x, y, z, u, v, w represent any variables.

— The auxiliary signs (,), ,.

We write $A1 = \Sigma 1 \cup \{\perp, \top, \neg, \wedge, \vee, \rightarrow, \forall, \exists, =, (,), ,\} \cup Var$

The signature $\Sigma 1$ is different in each alphabet A1 and may be finite or infinite, but we will always suppose that it is decidable.

Rules for the Construction of Terms and Formulas

Both the terms and the formulas of predicate logic are formal objects. The terms represent elements of a set that we will call domain of discourse. The formulas represent propositions concerning elements of a given domain of discourse. The domain of discourse is the formal object about which the first-order predicate logic sentences affirm or deny something.

Terms

Given an alphabet A1, we construct the set T of the terms of L1 through a rule for the definition of atomic terms (AtT) and another rule for the construction of the compound terms (CpT):

(AtT)   Each variable x of Var and each constant c
        of $\Sigma 1$ is an atomic term.

(CpT)   If f is a function of F with n arguments and t_1,
        ..., t_n are terms of T, then $f(t_1, \ldots, t_n)$
        is a compound term of T.

Example

Given the $\Sigma 1$ signature $\Sigma ar = \{0, s, +, \times, <\}$, it is possible to construct formulas of arithmetic.

The elements of Σar are

— The constant 0

— The successor function s with an argument

— The functions $+$ and $\times$ with two arguments

— The relation $<$ with two arguments

Given the signature $\Sigma ar = \{0, s, +, \times, <\}$ we can construct the term of arithmetic $(0 \times s(x))$ in four steps:

Step	Constructed	Rule	Terms or formulas used
1	x	(AtT)	
2	0	(AtT)	
3	s(x)	(CpT)	1
4	(0 × s(x))	(CpT)	2, 3

Formulas

Given an alphabet A1, we construct the set L1 of the first-order logic formulas through rules (AtF), ($\neg$F), and ($\square$F):

(AtF)   Both signs $\top$, $\bot$ as any sign p of proposition of Σ1 are atomic formulas of L1.
          If s,t $\in$ T, then (s = t) is a formula of L1.
          If P $\in$ P has n arguments and $t_1, \ldots, t_n \in$ T, then $P(t_1, \ldots, t_n)$ is a formula of L1.

($\neg$F)   If φ is a formula of L1, then $\neg\varphi$ is also a formula of L1.

($\square$F)   If φ1 and φ2 are formulas of L1, then $(\varphi_1 \square \varphi_2)$ is also a formula of L1.

($\forall,\exists$F)   If φ is a formula of L1, then $\forall x\varphi$ and $\exists x\varphi$ are also formulas of L1.

Examples

Supposing that Σar = $\{0, s, +, \times, <\}$, through the rules (AtF), ($\neg$F), and ($\square$F) we can construct, for example, the formula (s(x) = y) in four steps, and the formula $\forall z \neg((z \times z) = s(y))$ in seven steps:

Step	Constructed	Rule	Terms or formulas used
1	x	(AtT)	
2	y	(AtT)	
3	s(x)	(CpT)	1
4	(s(x) = y)	(AtF)	2,3

Step	Constructed	Rule	Terms or formulas used
1	z	(AtT)	
2	y	(AtT)	
3	s(y)	(CpT)	2
4	(z × z)	(CpT)	1
5	((z × z) = s(y))	(AtF)	3,4
6	¬((z × z) = s(y))	(¬F)	5
7	∀z ¬((z × z) = s(y))	(∀,∃F)	6

APPENDIX 4
Semantics of the First-Order Logic

THE SEMANTICS of first-order logic studies the possible formal meaning of the terms t of T and of the formulas ϕ of L1.

The possible formal meanings of a term t of T are elements of the domain of discourse and are defined through construction rules.

The possible formal meanings of a formula ϕ of L1 are true (which we represent with the sign 1) or false (which we represent with the sign 0). The formal meaning (true or false) of a formula ϕ of L1 is defined with the help of a Σ1-interpretation of this formula ϕ in a domain of discourse. The domain of discourse is a set of elements about which the first-order predicate logic formulas affirm or deny something.

Σ1-interpretations

A Σ1-interpretation (S,σ) of a formula ϕ is composed of two elements: A Σ1-structure S interpreting the signs of ϕ and a valuation σ of the variables of ϕ.

A Σ1 structure $S = (D; c^D \ldots; f^D \ldots; P^D \ldots)$ is given by

— A domain of discourse D, which is a set which is a not empty set.

— A meaning in D of each sign of the constant c of Σ1, which is an element of D denoted by c^D.

— A meaning in D of each function sign of f of Σ1 denoted by f^D, which is a list of elements of D, such that the last element of each list is the image of the previous n elements.

A meaning in D of each predicate sign P with n arguments of $\Sigma 1$ denoted by P^D, which is a set of ordered successions of n elements of D.

Given that a $\Sigma 1$ structure S and a set of variables Var, a valuation σ to the variables of Var is an application of an element $\sigma(x)$ of the domain of discourse of the structure S to each variable x of Var.

Example of a $\Sigma 1$-structure S

Given the signature $\Sigma ar = \{0, s, +, \times, <\}$ of the previous appendix, a possible Σar-structure is the following: $S = (N; 0^N; s^N; +^N; \times^N; <^N)$, where the domain of discourse is the set N of the natural numbers and the meanings of the symbols of the signature Σar are the usual ones:

0^N = the number 0

s^N = the successor function of N, which applies each natural number n to its successor $n + 1$

$+^N$ = the sum of N

$\times^N$ = the product of N

$<^N$ = the relation "less than" in N

Interpretation of Terms

Given a $\Sigma 1$ structure S and a valuation σ, we define a $\Sigma 1$-Interpretation (S,σ) by a function applying an element $[t]$ of the domain of discourse D to each term t of T.

The element $[t]$ is constructed as regards the structure of t as follows:

Base of the Definition

We define the semantic value of the atomic terms:

(AtT): $[x] = \sigma(x)$ for each variable x of Var

$[c] = c^D$ for each constant c of $\Sigma 1$

Definition Steps

(CpT): $[f(t1, \ldots, tn)] = f^D([t_1], \ldots, [t_n])$ for each sign of function f of $\Sigma 1$

Interpretation of Formulas

In order to define an interpretation of a formula φ we must previously define a modified valuation. Given a valuation σ of elements of the domain of discourse D of A to the variables of Var and given an element "a" of D, we call valuation σ modified in x, and we designate $\sigma[x/a]$, to a new valuation which coincides with σ in the element which assigns to all the variables except in the case of the variable x, which is assigned the element "a".

Now, for each Interpretation (S,σ) we define an application $[\ ]$ which makes each formula φ of L1 correspond to its semantic value $[\varphi]$, which can be 0 or 1.

$[\ ]: L1 \to \{0,1\}$

Base of the Definition

We define the semantic value of the atomic formulas:

(AtF): $[\top] = 1$

$[\bot] = 0$

$[s = t] = 1,$ if $[s] = [t]$

$[s = t] = 0,$ if $[s] \neq [t]$

$[P(t_1, \ldots, t_n)] = 1,$ if $P^D([t_1], \ldots [t_n])$

$[P(t_1, \ldots, t_n)] = 0,$ if not $P^D([t1], \ldots, [tn])$

Definition Steps

To define the compound formulas we follow these construction rules:

($\neg$F): $[\neg\varphi] = v_\neg([\varphi])$

($\Box$F): $[\varphi_1 \Box \varphi_2] = v\Box([\varphi_1], [\varphi_2])$

($\forall, \exists$F) $[\forall x \varphi] = 1,$ if $[\varphi] = 1$ for all possible valuations $\sigma[x/a]$

    $[\forall x \varphi] = 0,$ if $[\varphi] = 0$ for a valuation $\sigma[x/a]$

$$[\exists x\, \varphi]A = 1, \qquad \text{if } [\varphi] = 1 \text{ for a valuation } \sigma[x/a]$$

$$[\exists x\, \varphi]A = 0, \qquad \text{if } [\varphi] = 0 \text{ for all possible valuations } \sigma[x/a]$$

Models

Given a formula φ of L1 and an Σ1-Interpretation (A,σ),

If $[\varphi] = 1$ we will say that (A,σ) is a model of the formula φ

If $[\varphi] = 0$ we will say that (A,σ) is not a model of the formula φ

APPENDIX 5
Numerical Systems: Their Role in First-Order Logic

As NOTED in Appendixes 1 and 2, the language L0 allows us to analyze mathematical statements only to the level of depth of atomic statements. In this section, we begin to see how the language L1 is able to analyze mathematical statements beyond the atomic statements. Our first step in this process will be to understand the numerical systems that exemplify the domains of discourse. In turn, the domains of discourse contain the functions and relations—such as addition and multiplication—between specific elements that are contained within mathematical statements. As we look at examples of language L1 in Appendixes 3, 4, 7, and 10, the numerical systems and the domains of discourse will be our guide.

In classical mathematics, the numerical systems contain the most common domains of discourse, and the natural numbers are the most basic numerical system. We begin with them.

Natural Numbers
When we count objects, we attribute natural numbers to them: 1, 2, 3, and so on, passing from one number to its successor. The decimal representation of the numbers we know is such that we can always write the successor of any given number. Some authors establish the number 1 as the first natural number. We establish 0 as the first natural number and denote $N = \{0, 1, 2, 3, \text{etc.}\}$ as the set of natural numbers. The letters n, m, l serve to designate an arbitrary

natural number. The successor of number n we designate through s(n). That is to say, s(n) = n + 1.

One of the most notable properties of the natural numbers is the simplicity with which they can be generated from zero through the repeated application of the successor function. This facility to generate all the elements of a set with the help of an initial generator, in this case 0, and a method to obtain any other number starting from the generator, enables us to prove that all the members of this set, in this case the natural numbers, comply with a certain property P, provided that 0 complies with this property, and provided that, if any number complies with this property, its successor also complies with it.

Graphically we can represent N as follows:

0	1	2	3
o ———————— o ——————— o ———————— o - - - - -
0	s(0)	s(s(0))	s(s(s(0)))

Where: $1 = s(0), 2 = s(s(0)), 3 = s(s(s(0))), \ldots$

The natural numbers are a set ordered through the relation equal or less than. Given two natural numbers m, n, we say that m is equal or less than n and we write $m \leq n$, if both are equal or we can obtain n from m through the repeated application of the successor function. If $m \leq n$ and $m \neq n$, we will write $m < n$.

The set of natural numbers is "closed" as regards the sum—that is to say, given two natural numbers, their sum is always another natural number.

Integers

In order to resolve the equation $m - n = x$ when $m < n$, and both are natural numbers, we have to subtract a natural number from a smaller natural number and we obtain a negative number that is not natural. In order to obtain not only the result of the addi-

tion, but also the subtraction, we refer to the set of integer numbers
$Z = \{\ldots -2, -1, 0, 1, 2, \ldots\}$.

Note that, in the case of integer numbers the complexity of the
structure is increased, we cannot generate this only through the
succession function, but we have to add another generator function which we will call predecessor and is defined as $p(n) = n - 1$,
for any integer number n.

The structure of the integer numbers is a little more complex
than those of the natural numbers, although the property that
every integer number can be obtained starting from 0 continues to
be verified by repeatedly applying one of the two generator functions.

$$\begin{array}{ccccccc} -3 & -2 & -1 & 0 & 1 & 2 & 3 \end{array}$$

$$\begin{array}{ccccccc} p(p(p(0))) & p(p(0)) & p(0) & 0 & s(0) & s(s(0)) & s(s(s(0))) \end{array}$$

The product of two integer numbers m, n which we write as
$m \times n$ is an integer number.

Rational Numbers

The equation $x \times n = m$, $n \neq 0$ does not have an integer solution for
x, in the cases in which m is not a multiple of n, but a quotient of
integer numbers which we call a rational number.

By applying a magnifying glass to the segment that represents
the numbers between 0 and 1, we can represent some rational
numbers more intuitively:

$$\begin{array}{ccccc} 0 & \frac{1}{8} & \frac{1}{4} & \frac{1}{2} & 1 \end{array}$$

The set of rational numbers is represented by the letter Q and its
elements by q_1, q_2, q_3, etc.

Each rational number admits many equivalent representations, for example,

$$\tfrac{1}{2} = \tfrac{2}{4} = \tfrac{4}{8} = \cdots$$

Every rational number can be represented in decimal form. For example, $\tfrac{1}{3} = 0.333. \ldots$ The decimal representation of a rational number either has a finite number of figures such as $\tfrac{1}{5} = 0.2$, or gives recurring decimal values such as $\tfrac{1}{7} = 0.142857142857. \ldots$ The reason for recurrence is that by obtaining the decimals the rest of the division will always be less than the divider. If the divisor is n, there can only be n − 1 different subtractions, and when the subtraction is repeated, the period of decimals will be repeated.

Every rational number admits a representation with a positive denominator, because $\tfrac{m}{-n} = \tfrac{-m}{n}$. This fact enables us to define an ≤ order relation between the rational numbers as follows:

$$\tfrac{m_1}{n_1} \leq \tfrac{m_2}{n_2}$$

on the condition that $\tfrac{m_1}{n_1}, \tfrac{m_2}{n_2}$ are two representations of the same number with a positive denominator and $m_1 \times n_2 \leq m_2 \times n_1.$

Real Numbers

We have seen that, once the length of a unit segment is fixed, the numbers of the systems N, Z, and Q can be represented as the length of the segments of a straight line. We can present the problem inversely. Once the length of a unit segment is fixed, can the length of any segment be measured through a number of one of the N, Z, Q systems?

The answer is no. $\sqrt{2}$ cannot be represented as the quotient of two integer numbers; therefore, the diagonal and the side of a square are incommensurable—that is to say, they do not have a common measure.

Any decimal number $a, a_1 \ldots a_n$ can be written as a quotient $\frac{aa_1 \ldots a_n}{10\ldots0}$ where $10\ldots0$ has n zeros. If $\sqrt{2}$ were written as the quotient of two whole numbers, say $\sqrt{2} = \frac{m}{n}$, then $2n^2 = m^2$ and the factor 2 would appear an uneven number of times on the left-hand side of the equation and an even number of times on the right-hand side.

Therefore, the equation $\sqrt{2} = \frac{m}{n}$ does not have a rational solution with integer values for m and n, but there are algorithms that make it possible to approximate their values through recurring calculations of successive decimals. These algorithms cannot manage to calculate the exact value of $\sqrt{2}$ through a finite number of figures and cannot recur the successions of decimal values that they obtain since, in both cases, if this occurred, there would be two integer values m, n such that $\sqrt{2} = \frac{m}{n}$.

We refer to the numbers that can be represented by an integer number followed by a finite or infinite succession of decimals, regardless of whether these are recurring or not, as real numbers, and the set of these numbers is designated as letter R. Among these in addition to $\sqrt{2} = 1.4142\ldots$ are other numbers that frequently appear in mathematics, such as $e = 2.7182\ldots$ and $\pi = 3.141592\ldots$ and which cannot be represented as the quotient of two integer numbers.

By applying a microscope with infinite power that amplifies a segment as much as we wish, we can see how we can approximate to a real number that we will call an irrational number, through a succession of rational numbers:

Any real number can be constructed as an infinite succession of rational numbers. Thus, given a real number r, we can construct a succession of rational numbers as near to r as we wish, using a sufficiently large number of decimals.

Among their data, computers use integers and real numbers. Given that a computer can only store a finite amount of data, the numbers used by the computer only have a finite number of decimals, and these are approximations of real numbers that are always inexact in the case of irrational numbers.

We say that a real number r is algebraic when it is the solution to a polynomial equation of the $a_0x^n + a_1x^{n-1} + \ldots + a_n = 0$ type, where $a_0 \neq 0, a_1, \ldots, a_n$ are integer numbers and $n > 0$. If r is not algebraic, we say that it is transcendent.

If $q = \frac{m}{n}$ is a rational number, then q is algebraic as it is the solution to the polynomial equation $nx - m = 0$.

However, it can be proved that the real numbers $e = 2.7182 \ldots$ and $\pi = 3.141592 \ldots$ are transcendent.

The Complex Numbers

Continuing with the work involved in completing the solutions of the polynomial equations, we find that, although all odd-grade polynomials with real coefficients have a solution in the set R of the real numbers, the polynomial $x^2 + 1 = 0$ has no real solution as there is no real squared number worth −1. If we imagine the existence of a new nonreal number which we call *i*, we can deduce the set C of the complex number, in the form *a + bi*, where *a* and *b* are real numbers.

We can represent the complex numbers in a plane called the complex plane.

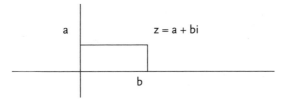

The fundamental theorem of algebra shows that the algebraic process that we have followed and which has led us to successive extensions of the sets of numbers:

$$N \rightarrow Z \rightarrow Q \rightarrow \text{Algebraic real numbers} \rightarrow C$$

finishes in the set C. In fact, this theorem states that in C there is not only a solution to the polynomial $x^2 + 1 = 0$, but that every polynomial with coefficients in C has a solution in C.

APPENDIX 6
The Paradox of Russell

THE CASE OF "Russell's Paradox" illustrates how mathematicians challenge each other's work, leading to new developments in mathematics. The history and content of Bertrand Russell's challenge to the work of Gottlob Frege follows.

In 1893, Frege published the first volume of the *Grundgesetze der Arithmetik*, and in 1903 he published the second volume. In the *Grundgesetze* Frege wanted to derive the mathematical theorems of logical laws and principles, but to do so, he introduced a logical principle that turned out to be inconsistent. Frege established a principle that we can call the principle of abstraction, according to which, given a well-defined property P, we can form the set C of all the elements X that have this property. We can write this definition more formally like this:

$$C = \{X| \ P(X)\}$$

In 1903, just before the second volume of the *Grundgesetze* appeared, Bertrand Russell wrote to Frege showing him that a contradiction was deduced from the principle of abstraction as it appeared in the *Grundgesetze*.

In fact, the following contradiction is deduced from the principle of abstraction: the property of a set X not to be an element of itself is well defined in the language of Frege. We can write the property formally: $X \notin X$, where "$A \in B$" means "A is an element of the set B" and "$A \notin B$" means "A is not an element of B." Then, by the principle of abstraction we can form the following set:

$$C = \{X \mid X \notin X\}$$

In a less formal language we would say: C is the set whose elements are all the X such that X is not an element of X $(X \notin X)$

We can now ask, is C an element of C? Or with formal notation, $C \in C$?

The response is paradoxical as

if $C \in C$, then $C \notin C$
and if $C \notin C$, then $C \in C$

One way to resolve the paradox of Russell is to substitute the principle of abstraction with the principle of comprehension.

The principle of comprehension states that, given a set X and a well-defined property P, we can form the subset S of X—written $S \subseteq X$—made up of all the elements Y of X that comply with the property P. With our notation

$$S = \{Y \mid Y \in X, P(Y)\}$$

That is to say: S is the subset of all the elements Y of X such that Y has the property P.

According to the principle of comprehension, the property of not being an element of itself ceases to be paradoxical.

In fact, based on a previous set X, by applying the principle of comprehension, we obtain the subset $S \subseteq X$:

$$S = \{Y \mid Y \in X, Y \notin Y\}$$

In less formal language, S is the subset of all the elements Y of X such that Y is not an element of Y $(Y \notin Y)$, and in this case $S \notin X$ does not imply $S \in X$.

The Completeness of First-Order Logic

THE ARGUMENTATION of L0 (and of L1) consists of a finite sequence of declarative propositions called premises from which follows another declarative proposition called the conclusion. The argumentation of logic L0 (and L1) can be studied semantically or syntactically. Semantically we study the validity of argumentations in accordance with the semantic meaning of their premises and conclusion. Syntactically we study the validity according to some formal calculus of deduction.

Semantically an argument of L1 is valid if every interpretation that is a model of the premises is also a model of the conclusion. Syntactically an argument of logic L1 is valid in a certain formal calculus if it meets the formal rules of inference peculiar to that calculus.

Quite simply, then, an argument in formal logic is valid when the premises are taken to be true and, based on that, the conclusion is also found to be true. Here is an example of a formally valid argument: In the diagram below, we affirm that argument A is formally correct without knowing the meaning of the (fictional) words *zopiloid* and *pombiform*. That is because, on condition that the words "zopiloid" and "pombiform" have meanings that make the premises true, that same meaning will make the conclusion also true. Moreover, we can affirm this thanks to the syntactic structure of the premise and the conclusion.

A   All zopiloids are pombiforms   [premise]
     Arthur is zopiloid   [premise]

     ∴ Arthur is pombiform   [conclusion]

However, the following argumentation is not formally valid:

B   Some calorts are equalized   [premise]
     John is a calort   [premise]

     ∴ John is equalized   [conclusion]

Argumentation B is not valid because the outcome can be that the premises are true and the conclusion false. In fact, regardless of the meaning of calort and equalized we can affirm that, based on the syntactic structure of the premises, although some calorts are equalized and John is a calort, John may be a calort who is not equalized.

The completeness theorem of the L1 logic, that Gödel proved in his doctoral thesis, establishes that there is some formal calculus by which we can infer all arguments semantically valid in L1. We now know various formal calculi that allow full deduction of all arguments of L1 that are semantically valid.

The significance of Gödel's completeness theorem is the fact that Gödel proved that there is an effective formal calculation by which we can prove that a given argument is semantically valid.

APPENDIX 8
Jack's Formula

As WE SEE in the story of six friends, Jack does not have to select a valuation for the only variable x that appears in φ because this variable appears as affected by the quantifier ∀ and the meaning of φ will be the same for any valuation that we make for x. Thus, instead of referring Jack's formula to an interpretation $(A,φ)$, it is sufficient to refer it to a structure.

Therefore, in writing his formula φ on the blackboard— $φ:∀x\,((R(x) → SL(s,x)) ∧ (F(x) → DR(s,x)) ∧ (B(x) → FE(s,x)) ∧ (KF(s,x) → (M(x) ∨ W(x))))$—Jack made particular choices: First of all, we see that Jack selected the signature

$$Jk = \{s, R, SL, F, DR, B, FE, KF, M, W\}$$

where "s" is a sign of a constant that formalizes the word "Spirit"; $R(x), F(x), B(x), M(x), W(x)$ are predicate signs with an argument that respectively formalize the propositions: x is a river, x is a flower, x is a bird, x is a man, and x is a woman; $SL(x,y), DR(x,y), FE(x,y), KF(x,y)$ are predicate signs with two arguments that respectively formalize the propositions: x sleeps in y, x dreams in y, x feels in y, x knows y.

As φ is a formula of first order, we can represent its models as ΣJk-structures of the form:

$$S = (D; s^D; R^D, DL^D, F^D, DR^D, B^D, FE^D, KF^D, M^D, W^D)$$

Where the domain of discourse D is any set other than an empty set, s^D is an element of A that represents the meaning of the sign s in the Σ-structure S; R^D, F^D, B^D, M^D and W^D are subsets of D that represent the meaning, in the Σ-structure S, of the predicate signs R, F, B, M, and W; SL^D, DR^D, FE^D, and KF^D are sets of ordered pairs <x,y> of elements of D that represent the meaning, in the Σ-structure A, of the predicate signs SL, DR, FE, and KF.

The semantic meaning of the formula that Jack wrote on the blackboard is not determined. Jack did not specify the semantics of the formula he wrote. The formula Jack wrote can be interpreted in different models. We do not even know in which domain of discourse this has to be interpreted.

APPENDIX 9
Maria's Formula

In this story, Jack did not specify the semantics of the formula he wrote on the blackboard, so his formula is not determined. His formula can be interpreted in different models. We do not even know in which domain of discourse this has to be interpreted.

This enables Maria, the mathematician, to give an interpretation to Jack's formulation, which is particular although arbitrary. Maria specifies the domain of discourse D determining that D is the set Z of the integer numbers. Maria also specifies the meaning of the signs Jack used. According to Maria's specification, s^Z is a component of the set Z; in particular, it is the number 4. R^Z, F^Z, B^Z, M^Z, and W^Z are subsets of Z formed respectively by numbers greater than 4, greater than 5, greater than 6, by the number 2, and by the number –2, and finally SL^Z, DR^Z, FE^Z, and KF^Z are subsets of Z formed respectively by the pairs <x,y> such that "x is less than y," "x is less than y – 1," "x is less than y – 2," "x is the square root of y."

Example of an L1 Formalization

LET US SUPPOSE that we wish to formalize the following mathematical proposition in first-order predicate logic: "x + 1 is not the square of a natural number." The proposition "x + 1 is not the square of a natural number" is couched in an academic language that is a mixture of a natural language, in this case English, with some formalisms. In order to formalize this proposition, we take the following steps:

We take the Σar-structure $S = (N; 0^N; s^N; +^N; \times^N; <^N)$, which we have seen in Appendix 4 as suitable for interpreting this proposition, where the domain of discourse of S is the set N of natural numbers.

We assign an arbitrary value to the variable x. For example $\sigma(x) = 5$ or, for example, $\sigma(x) = 3$.

We formalize the meaning of the words that appear in the proposition "x + 1 is not the square of a natural number" by using signs of first-order predicate logic and signs of the signature Σar.

"For every y $(\forall y)$, it is false that $(\neg)$ the product of y by y $(y \times y)$ is equal to $(=)$ x plus the successor of 0 $(x + s(0))$."

We construct a formal proposition $\forall y \neg (y \times y = x + s\,(0))$ and interpret it in the formal structure S.

Now, we can easily verify that if σ assigns, for example, the value 5 to x, then the interpretation (S,σ) is a model of the formula φ. In fact, for each natural number n, it is false that its square is 5 + 1.

Likewise we can verify that if σ assigns the value 3 to x, then the interpretation (S,σ) is not a model of the formula φ as there is a natural number whose square is $3 + 1$.

In both cases we can verify that the meaning of φ in the interpretation (S,σ) coincides with the meaning that is usually given to the proposition "x + 1 is not the square of a natural number" in academic mathematical language, when this proposition is taken as an affirmation concerning natural numbers.

⚘ Glossary

This glossary is designed as a guide for this particular book and is therefore far simpler than found in formal mathematics textbooks. Readers may refer to university and professional sources for more precise and in-depth definitions.

algebra   Elementary algebra is a branch of mathematics that substitutes letters for numbers. It is used to solve polynomial equations. An algebraic equation represents a scale, so that what is done on one side of the scale with a number is also done on the other side of the scale. Modern algebra uses more abstract methods for advanced problems.

algorithm   A step-by-step problem-solving procedure for solving computational mathematical problems. Not to be confused with logarithm.

arithmetic   A branch of mathematics usually concerned with the four operations (adding, subtracting, multiplication, and division) of positive numbers.

attractor   The direction in which the elements of a dynamic system tend to move. The attractor may be a point, curve, or manifold. It is a feature of chaotic systems in which order emerges from apparent chaos.

axiom   A statement that needs no proof because the truth is obvious. This kind of self-evident proposition is also called an intuitive truth. A postulate is the same as an axiom.

base number   A whole number made the fundamental number and raised various powers to produce the major counting units of a number system. For example, 10 is the base number for Western counting and measurement, while 60 is the base number for counting time.

calculus   A method of calculation using a special system of notation in symbols. This system of mathematical analysis combines differential calculus and integral calculus. It is often used to describe changing values of objects and spaces in motion.

classical mathematics   Based on the logic of Aristotle, this approach to mathematics allows proofs by finding contradictions in opposites. It also accepts infinite sets. Classical mathematics often studies systems based in axioms like the postulates of Euclid.

coefficient   A factor that multiplies a related numerical factor. In the equation $2x = 10$, the coefficient of x is 2, and vice versa.

completeness   The ability of a mathematical system to find basic axioms and theorems that will prove true or false all of its parts in one consistent whole. The discovery that this is not possible in arithmetic gave rise to the concept of incompleteness in mathematics.

complex numbers   An ordered pair of numbers that includes a real number and an imaginary number. The system was first used to solve special cubic and quartic equations. They also allowed mathematicians to speak of the square root of a negative number.

composite numbers   All natural (counting) numbers that are not prime numbers.

conjecture   In mathematics, a proposed theory about the relationships of certain kinds of numbers that seems universally true according to a limited number of calculations, but cannot be ultimately proved because all the calculations necessary have no apparent limit.

consistent   A set of mathematical statements is consistent when we cannot deduce from it a contradictory statement.

constructivist mathematics   A mathematical method and school that arose in response to the paradoxes found in classical mathematics. It limits mathematics to direct proofs, and it does not accept the idea of actual infinite sets. The term suggests that mathematics is "constructed" by the human mind for specific and limited calculations, but has no absolute or infinite status in reality.

decidable set   A set is decidable if we have an effective criterion to recognize whether a given sign belongs or does not belong to it.

deductive reasoning   Reasoning from a known principle to an unknown, from the general to the specific, or from a premise to a logical conclusion. Mathematics is typically seen as the classic deductive science.

divisor   The number that is doing the dividing. This number is found outside of the division bracket. In $\frac{a}{b} = c$, b is the divisor.

domain of discourse   Contains all the elements giving a meaning to a mathematical formula. The elements of a domain of discourse may be numbers, points, planes, or any mathematical object.

empirical language   The language of natural science, which combines abstracted signs (such as $E = mc^2$) with descriptions and measurement of physical objects and forces perceived by the senses.

equation   A statement showing the equality of two expressions usually separated by left and right signs and joined by an equal sign.

Euclidean geometry   The mathematical principles that the Greek thinker Euclid proposed for three-dimensional space, which includes points, lines, planes, angles, and three-dimensional objects. Euclidean geometry defined Western mathematics for two thousand years and remains the basis for everyday geometry.

even number   A natural (counting) number that can be divided by 2.

first-order predicate logic   A fundamental system of logic that builds upon and extends propositional logic by defining a domain of discourse for that simpler logic, and by defining functions and relations on the elements of this domain. First-order predicate logic is used in mathematics, computer science, and philosophy.

formal language   A language consisting of well-defined sequences of signs rather than ordinary words and typified by logic, mathematics, and computer science.

formula   A rule that describes the relationship of two or more elements that may be constants or variables. An equation stating the rule.

geometry   A field of mathematics that studies lines, angles, shapes, and their properties. Geometry is concerned with physical shapes and the dimensions of the objects.

incompleteness   The idea that all but the simplest axioms in arithmetic have a limited power of explanation, thus making arithmetic not complete as a whole. In other words, any theorem that can prove some truth in arithmetic cannot prove all truths in arithmetic.

inductive reasoning   Reasoning that moves from particular facts or individual cases to general conclusions. It is the opposite of deductive reasoning and is typified by the empirical sciences.

infinite   Lacking limits or bounds in space, quantity, or time. In math, infinite means indefinitely large: greater than any finite number, however large. The infinite is capable of being put into one-to-one correspondence with a part of itself (this is an infinite set).

infinitesimal calculus   The method of mathematical analysis of differential and integral calculus. Infinitesimal means too small to be measured, so in mathematics such a number is said to be smaller than an already-assigned nonzero number in a calculation.

integers   A set that includes all natural (counting) numbers, the negatives of all natural numbers, and also zero.

intuition   The direct knowing or learning of something without conscious reasoning. An immediate apprehension or understanding.

irrational number   A number that cannot be represented as a decimal or as a fraction. A number like pi is irrational because it contains an infinite number of digits that keep repeating. Many square roots are irrational numbers.

logic   Sound reasoning and the formal laws of reasoning.

mathematics   A field made up of four traditional branches: arithmetic, geometry, algebra, and analysis. Analysis is concerned with estimates, inequalities, differential and integral calculus, and properties of the real numbers.

metalanguage   Any language used to analyze or summarize another language.

metamathematics   The application of mathematical language, principles, and analysis to understand a mathematical system or the relationship between different mathematical systems.

metaphysical language   Any language that addresses questions of first and absolute principles. It typically uses symbols shaped by various cultures to speak of the origins and fate of individuals and the universe. Terms such as "Prime Mover," "Creator," "eternal life," and "free will" are symbols used in this kind of philosophical and religious language.

metaphysics   A branch of philosophy that looks at first principles and seeks to explain the nature of being or reality (ontology) and the origin and structure of the universe (cosmology). Meaning "beyond physics," the term was applied to the section that came after Aristotle's *Physics* when his works were collected.

mixed numbers   Mixed numbers refer to whole numbers with fractions or decimals. Examples are $3\frac{1}{2}$ or 3.5.

modal logic   A method of logical analysis that looks at what is possible, impossible, and necessary in logical systems. Normal logic limits itself to finding whether any or all elements of the domain of discourse satisfy a certain property.

myth   A term that explains the use of religious stories and narrative, usually about the people in one culture or another, that does not connote falsehood, but rather the combining of facts with ultimate principles and beliefs. Examples are the so-called creation myths, nonscientific explanations of the origins of the universe that can have parallels to scientific explanations.

natural language   The human language of different cultures, races, ethnicities, and nations. This kind of language is the logical basis for mathematical language. Empirical science joins natural language with mathematics into an academic language common in research today.

natural numbers   Regular counting numbers. They include 1, 2, 3, and so on, and are denoted as N = {0, 1, 2, 3, etc.}. The number 0 was not accepted into a numeral system until the ninth century.

negative number   A real number that is less than zero. This can include a decimal such as .10 as well as include −2, −3, −4, and so on.

non-Euclidean geometry   A geometry that rejects any of Euclid's postulates, especially the postulate that mandates parallel lines in space. Non-Euclidean geometry is used to explain the shape of curved space as theorized in Einstein's theory of relativity regarding gravity.

odd number   A whole number that is not divisible by 2.

ontological argument   The logical argument that it is necessary for God to exist—God being defined as the greatest that can be conceived—because it is "greatest" to exist in reality and thought than to exist only in thought. This form of the argument uses reduction to the absurd logic (it would be a contradiction for the greatest to be less than the greatest). Another version says that since existence is part of perfection, a perfect being must exist.

operation   Any arithmetical action of addition, subtraction, multiplication, or division.

paradox   In mathematics, a statement that is self-contradictory and therefore false.

Platonism   A philosophical school following Plato that believes that all things, including mathematics and numbers, have absolute Forms (or Ideas) beyond the transient world of the physical senses. The rational mind, or intuition, is believed capable of apprehending these perfect Forms.

polynomial equation   An algebraic equation consisting of several terms. A monomial equation has a single term. Polynomials include variables as well as more than one term.

positive number   A real number greater than zero.

positivism   A modern scientific philosophy that holds that the only valid and reliable human knowledge is based upon physical objects that can be quantified and named precisely. This is commonly called philosophical materialism or metaphysical naturalism.

predicate   In logic as in grammar, an assertion or description made about the subject of a proposition. To say that the grass is green is to make "green" a predicate of the subject, which is the grass. In logic, a predicate is either affirmed or denied about the subject in a proposition.

prime numbers   A number that is divisible only by itself and 1.

probability   In mathematics, the number of times something will occur over the range of possible occurrences, expressed as a ratio. Probability is an important aspect of statistics and the study of chaotic systems, whose systems have probably but not deterministic patterns.

propositional logic   The simplest form of logic, which is based on a string of single declarative propositions that can be judged true or false.

Pythagorean school   The followers of the early Greek mathematician Pythagoras. They believed the universe originated in a few important numbers, especially 10. This led to a kind of mystical religion, which is now extinct, but which has influenced mystical and mathematical thought to the present.

Pythagorean theorem   The theorem that predicts the exact relations of the three sides of any triangle with a right angle.

radius   The distance from the center of a circle or sphere to the outside edge.

ratio   The relation between two quantities. Ratios can be expressed in words, fractions, decimals, or percents.

rational number   A number that can be expressed as the ratio of two integers.

real numbers   The combined set of rational numbers and irrational numbers.

reduction to the absurd In Latin known as *reductio ad absurdum*, it means to affirm that a proposition is true by proving that its opposite is false. Classical mathematics allows this form of proof, but constructivist mathematics does not always allow it.

representational language The language of natural science that may use some mathematical signs but aims primarily to represent the physical shapes, dynamics, and relationships of objects in nature.

semantics The study of how the structure of a language conveys its meaning, especially in its context. In logic and mathematics, semantics looks at the way signs (mathematical notations) operate according to rules of syntax. In humanities, semantics (also called semiotics) looks for the way symbols and signs convey meaning that is not apparent on superficial observation.

sets Any collection of objects. The most common objects in mathematics are the numbers and the points, lines, and planes of geometry. Set theory has unified the semantics of modern mathematics.

signs Often taken to mean the same thing as "symbols," but in mathematics, it technically is a notation that stands for a number or function or relationship between numbers, and any kind of sets.

structure In mathematics, a set along with functions and relations defined within it.

symbols Often taken to mean the same as "sign," a symbol has a larger content able to speak of human experience, narrative, myth, or metaphysical meaning. Symbols often are taken from wider human perceptions (such as those of an animal or invisible spirit), not merely an abstracted sign (such as an x or a y).

syntax The arrangement of words in a sentence, defining particular ways they modify each other. In logic, syntax applies to language in the abstract with no meaning attached to either the signs and notations or the ordering of signs and notations.

theorem In mathematics, a proposition that is not self-evident and must be proved. Not to be confused with a scientific "theory" in the empirical sciences, which is considered valid after sufficient testing of the hypothesis that proposed the theory.

theory of systems   An analytical method that looks for an overall unity in nature (and mathematics) by detecting similar patterns in disparate systems. An example of a similar pattern is the ability of many systems to self-generate.

Turing machine   A theoretical device conceived by Alan Turing that can read a succession of signs on a strip of tape and produce a series of calculations. It is the basis of the logical algorithm in a computer. Turing called it "intelligent machinery" since its power is not constrained by the limitations of a normal calculator.

undecidability   The fact that some calculations cannot be decided as true or false. It also means that a formula may be decided in its own limited way, but cannot be decided in wider logical theory. Undecidability points to the pluralism, limits, and openness in many mathematical systems.

whole numbers   The set of numbers that includes zero and all of the natural numbers.

Essay on Sources

I HAVE REFERRED to mathematical and literary sources in this text, and I hope the following information for select chapters allows readers to explore the topics further.

Chapter 2

In discussing the ontological argument, I have quoted from Brian Davies and G. R. Evans, eds. *Anselm of Canterbury: The Major Works* (Oxford: Oxford University Press, 1998), 87. An English translation of Anselm may also be found at http://www.sacred-texts .com/chr/ans. I have quoted John Henry Newman from *Meditations and Devotions of the Late Cardinal Newman* (New York: Longman, Green, 1911), 301, and Ignatius of Loyola from *Saint Ignatius of Loyola: Personal Writings* (New York: Penguin Books, 1996), 289.

Chapter 4

An illustrated edition of Euclid's *Elements* is found at http://aleph0 .clarku.edu/~djoyce/java/elements/elements.html. Euclid's proof of the Pythagorean theorem, as seen in proposition 47 of book 1 of the *Elements*, is found at http://aleph0.clarku.edu/~djoyce/java/ elements/bookI/propI47.html. Examples of the use of the pentagram in history may be found at http://mathworld.wolfram.com/. I have quoted Aristotle regarding the Pythagoreans from *Metaphysics* in *Aristotle's Metaphysics*, trans. W. D. Ross (Oxford: Clarendon Press, 1924), book 1, chapter 5. This entire translation is available at

http://classics.mit.edu/Aristotle/metaphysics.1.i.html. A translation of Aristotle's *Physics* may be found at http://classics.mit.edu/
Aristotle/physics.html.

Chapter 5

In this chapter, we learned about Aristotle's four causes, challenged
by the scientific approach of Galileo. Here, I quote Aristotle's exact
definitions of the four causes as found in *Aristotle's Metaphysics*,
trans. W. D. Ross (Oxford: Clarendon Press, 1924), and also at
http://classics.mit.edu/Aristotle/metaphysics.5.v.html:

> "Cause" means (1) that from which, as immanent mate-
> rial, a thing comes into being, e.g. the bronze is the cause
> of the statue and the silver of the saucer, and so are the
> classes which include these. (2) The form or pattern,
> i.e. the definition of the essence, and the classes which
> include this (e.g. the ratio 2:1 and number in general are
> causes of the octave), and the parts included in the def-
> inition. (3) That from which the change or the resting
> from change first begins; e.g. the adviser is a cause of the
> action, and the father a cause of the child, and in general
> the maker a cause of the thing made and the change-
> producing of the changing. (4) The end, i.e. that for the
> sake of which a thing is; e.g. health is the cause of walk-
> ing. For "Why does one walk?" we say; "that one may be
> healthy"; and in speaking thus we think we have given
> the cause. The same is true of all the means that inter-
> vene before the end, when something else has put the
> process in motion, as e.g. thinning or purging or drugs
> or instruments intervene before health is reached; for
> all these are for the sake of the end, though they dif-
> fer from one another in that some are instruments and
> others are actions.

In this chapter also, Galileo's quote about nature as a "grand book" is from Galileo Galilei, *Il Saggiatore* (*The Assayer*) (Rome, 1623). The original is in Italian; an English translation is at http://www.princeton.edu/~hos/h291/assayer.htm. Gottfried Leibniz is quoted from G. W. Leibniz, *Discourse on Metaphysics*, which may be found at http://www.anselm.edu/homepage/dbanach/Leibniz-Discourse.htm. I quote Leibniz on "let us make calculations" from his essay "The Art of Discovery," found in *Leibniz: Selections*, ed. Philip P. Wiener (New York: Charles Scribner's Sons, 1951), 51. The basic work of Pierre-Simon Laplace on a deterministic and mathematical universe is his *Philosophical Essay on Probabilities*, trans. and ed. A. I. Dale (New York: Springer-Verlag, 1998). Another version of this book is available at http://www.archive.org/details/philosophicaless00lapliala. David Hume's comment against metaphysics is found in chapter 12, part 3 of his *An Enquiry Concerning Human Understanding*. The Harvard Classics version of *An Enquiry* (New York: Collier and Son, 1909–1917) is at http://www.bartleby.com/37/3/.

Chapter 6

My references to Alonzo Church and Alan Turing may be found in Alonzo Church, "An Unsolvable Problem of Elementary Number Theory," *American Journal of Mathematics* 58 (1936): 345–63, and A. M. Turing, "On Computable Numbers, with an Application to the Entscheidungsproblem," *Proceedings of the London Mathematical Society*, ser. 2, 42 (1936–37): 230–65. Turing's 1947 quote is from A. M. Turing, "Lecture to the London Mathematical Society on 20 February 1947," typescript in the Turing Archive, King's College, Cambridge and can be found at http://turingarchive.org/browse.php/B/1.

Also in this chapter, I note that our belief in consistency in mathematics is a metaphysical presumption. To elaborate, I would like to add the following discussion: In 1936, Gerhard Gentzen (d.

1945), using transfinite induction from outside arithmetic, proved that arithmetic is consistent. However, as stated by E. B. Davies, in *Gödel, Inconsistency, Provability, and Truth: An Exchange of Letters* in Notices of the AMS, April 2006, 462–63, "If one proves the consistency of arithmetic by invoking some other, richer, formal system, one achieves nothing unless one considers that the consistency of that new system is less capable of being doubted." It is normal that the consistency of a more complex system than arithmetic is more questionable than the consistency of arithmetic. What would occur if arithmetic were inconsistent? Arithmetic is the most basic formal system of mathematics. If arithmetic were inconsistent, the arithmetical reasoning would only be valid as long as we do not find any inconsistencies in it. Thus, the value of arithmetical reasoning would cease to have *absolute* value and have only *local* value, it would have value only in its consistent subsystems. The mathematical derivations would continue to be valid in order to establish equivalencies between formulations, equations, and theories, and would also serve to predict properties of computer programs. Mathematics would maintain its current value on the condition that it is limited to mathematical arguments and computer processes within the limits in which the systems, or parts of the systems, maintain their consistency.

Chapter 8
In the story of six friends, I refer to their knowledge of the problems in mathematical completeness and decidability as revealed by the work of Gödel and Turing. To explain further, Gödel's First Incompleteness Theorem states that if Peano Arithmetic is consistent, then there is a formula G of Peano Arithmetic such that neither it, nor its negation, are demonstrable in Peano Arithmetic. Gödel's Second Incompleteness Theorem states that if Peano Arithmetic is consistent, then the formal proposition which expresses the consistency of Peano Arithmetic is not demonstrable within Peano Arith-

metic. The Halting Turing-Church Theorem states that there is no computable function capable of determining whether any computable function gives an output with any input.

Chapter 9

The letter by John Paul II to George V. Coyne is at http://www .vatican.va/holy_father/john_paul_ii/letters/1988/docu-ments/hf_jp-ii_let_19880601_padre-coyne_en.html. Stephen Jay Gould is quoted from Stephen Jay Gould, "Nonoverlapping Magisteria," *Natural History* 106 (March 1997): 16–22, and may also be found at http://www.stephenjaygould.org/library/ gould_noma.html.

In my closing comments on the role of the "great religions" in a scientific age, I would like to add this set of questions for the reader's reflection and further research by participants in the science-religion dialogue:

1. If the cosmologies of the ancient Near Eastern world could be purified and assimilated into the first chapters of Genesis, might not contemporary cosmology have something to offer to our reflections upon creation?

2. Does an evolutionary perspective bring any light to bear upon theological anthropology, the meaning of the human person as the *imago Dei*, the problem of Christology—and even upon the development of doctrine itself?

3. What, it any, are the eschatological implications of contemporary cosmology, especially in light of the vast future of our universe?

4. Can theological method fruitfully appropriate insights from scientific methodology and the philosophy of science?

Appendixes

The Appendixes 1, 2, 3, and 4 provide an elementary introduction to the syntax and semantics of classical Propositional and First-

Order Predicate Logic. There are many books on logic for non-mathematicians, among which the interested reader can choose the book *Logic: An Introduction to Elementary Logic* by Wilfrid Hodges (London: Penguin Group, 2001). In this book, the author manages to introduce most basic logical concepts without heavy mathematical machinery and notation.

Index

fifth postulate of, 72–73
geometry and, 5, 8, 10, 46, 71–74,
86, 124
Euclid of Megara, 49
Euclides ab Omni Naevo Vindicatus
(Euclid Preserved from Any
Error) (Saccheri), 72
Europe, 44, 53, 54, 63
excluded middle, principle of,
11–12

faith, 31, 32, 42, 126–27, 128, 129,
131
Feyerabend, Paul, 122
first-order predicate logic (L1),
27–31, 75, 78, 79, 85, 106, 113,
124
alphabet of, 139–40
completeness of, 156–57
formalization, 141–42, 161–62
interpretations, 143–44, 145
language of, 107–8, 158–59
models, 146
numerical systems and, 147–53
semantics of, 143–46, 156–57
structure, 144
syntax of, 98, 139–42, 156–57
terms of, 140–41, 143, 144
forgiveness, 32
formal signs, language of, 4, 5–6,
33, 42, 47, 49, 67, 69, 87, 89, 90,
114, 117
free will, 64
Frege, Gottlob, 4, 75–76, 83, 85,
93, 154

Galileo Galilei, 37, 55–61, 88,
114–15
geometry, 29, 36, 38, 48
Euclidean, 46, 71–74, 124
logic and, 77

non-Euclidean, 5, 8, 10, 46, 71, 73,
74, 86
God, 17, 18, 42, 63, 64, 127, 128, 131.
See also supreme being
formulation of ontological
argument for existence of,
27–31
ontological argument for
existence of, 23–27, 129
proof for existence of, 54
relationship with, 32
view of, 91
Gödel, Kurt, 24, 25, 49, 79, 80, 81,
82, 83, 84, 85, 111, 157
Goldbach, Christian, 83, 86
Gould, Stephen Jay, 129–30
gravity, 61–62, 74, 86, 114, 115
Greece
ancient, 36, 37, 44, 45, 49, 54, 127
mathematics/metaphysics in
ancient, 50–53
pre-ancient, 41, 42
Grundgesetze der Arithmetik
(Frege), 154
Grundlagen der Geometrie
(Foundations of Geometry)
(Hilbert), 73–74

harmony, 64, 130
Heisenberg, Werner, 9
Hermite, Charles, 77
heuristics, 90
Hilbert, David, 73–74, 77, 80, 81,
82, 84
Hinduism, 18, 42
Hippasus of Metapontum, 52
history, 20, 34
context of, 17
of mathematics, 35–37, 38
hope, 131
Hume, David, 65, 66, 116

interpretation of, 108–10
language of, 107–8
models of, 138
natural language and, 99–100
semantics of, 99–105, 136–38,
 156–57
syntax of, 98–99, 133–35, 156–57
propositions, 5, 11, 13, 28, 29, 33, 45,
 64, 75, 86, 89
atomic/compound, 95, 98, 102
calculus and, 49
false, 98
Proslogion (Anselm), 25
Ptolemy, 8, 59
purpose, human, 31–32
pyramid, volume of, 40–41
Pythagoras of Samos, 36, 41, 42,
 45, 50
Pythagorean Tetraktys, 51
Pythagorean theorem, about sides
 of triangle, 46
Pythagorean triples, 41–42
Pythagoreans, 17, 50–51, 115

qualia, 123
quantum physics, 9, 14, 22, 87, 121,
 123, 125, 127

al-Rashid, Harun, 53
rationality, 35, 37, 88, 91
reality, 3, 5, 7, 10, 15, 102, 109, 118,
 126, 130
logic and, 26
mathematics and, 52, 53, 87, 88
thought and, 29
reason, 4, 14, 35, 45, 86
sufficient, 63, 64
ultimate, 15
"reduction to absurd," 12, 98
relationship(s), 17–18, 76
causal, 57

with God, 32
between numerical amounts,
 39–40
religion(s), 15, 19, 44, 110, 112
Abrahamic, 31–32
communities and, 18
culture and, 128
infinity in, 42–43
logic and, 31, 33
magisteria of science and, 129–32
science and, 33, 59–61, 69, 105,
 126–29
symbols of, 25, 32, 60
Wicca, 51
representational models, 7–11
representational signs, language
 of, 5–6, 7, 22, 33, 68, 105
Republic (Plato), 53
resolution, lack of, 125, 126
Rhind, Henry, 41
Rhind papyrus, 41
risk, 90, 92
Roman Catholic church, 31, 56, 60,
 129, 130
rules, 29, 45
formal, of calculations, 47, 48
of inference, 49, 76
of logic, 4, 47, 48, 49
mechanical, 47, 48
Russell, Bertrand, 84
paradox of, 4, 77, 154–55

Saccheri, Girolamo, 72, 73
salvation, 19
science, 23. *See also* empirical
 science; natural science
applied, 86
autonomy of, 60
communities of, 9, 11, 13, 31, 122
computer, 39
instruments of, 88